Use and Abuse
of
America's Natural Resources

Use and Abuse of America's Natural Resources

THE

NATURAL LAWS OF HUSBANDRY

BY

JUSTUS VON LIEBIG

ARNO PRESS
A NEW YORK TIMES COMPANY
New York • 1972

Reprint Edition 1972 by Arno Press Inc.

Reprinted from a copy in The Princeton University Library

Use and Abuse of America's Natural Resources
ISBN for complete set: 0-405-04500-X
See last pages of this volume for titles.

Manufactured in the United States of America

Library of Congress Cataloging in Publication Data

Liebig, Justus, Freiherr von, 1803-1873.
The natural laws of husbandry.

(Use and abuse of America's natural resources)
Reprint of the 1863 ed.
1. Agricultural chemistry. 2. Agriculture.
3. Fertilizers and manures. I. Title. II. Series.
S585.L75 1972 631.8 72-2852
ISBN 0-405-04541-7

THE

NATURAL LAWS OF HUSBANDRY

BY

JUSTUS von LIEBIG

EDITED BY

JOHN BLYTH, M.D.,

PROFESSOR OF CHEMISTRY IN QUEEN'S COLLEGE, CORK.

NEW YORK:
D. APPLETON AND COMPANY,
443 & 445 BROADWAY.
1863.

EDITOR'S PREFACE.

IN the following work Baron Liebig has given to the public his mature views on agriculture, after sixteen years of experiments and reflection. The fundamental basis of the work is still the so-called Mineral Theory, which holds that the food of plants is of inorganic nature, and that every one of the elements of food must be present in a soil for the proper growth of a plant. The discovery of the remarkable power of absorption possessed by arable soils has necessarily led to a modification of the views regarding the mode in which plants take up their food from the soil. As the food of plants cannot exist for any length of time in solution in soils, it is clear that there cannot be a circulation of such solution towards the roots, but the latter must go in search of food. Hence the great importance of studying the ramification of the roots of plants, and the mode of growth of the different classes of plants cultivated by man. The first chapter is devoted to the consideretion of the growth of plants, of the formation of their roots, and of their power of selecting food, and the part played by the mineral matters which are absorbed.

If the food of plants is not in solution in the ground, we can conceive that those portions of the soil traversed by the numerous root ramifications will be more or less exhausted of food elements, whilst the immediate neighbouring portions are still rich in them. If, therefore, a succeeding crop is to grow equally well on all parts of a field, there must be a thorough mixing of the exhausted and of the unexhausted portions of soil. This is effected by mechanical means, by manures, or by certain chemical compounds. Hence the necessity of becoming acquainted with the nature and properties of the soil and subsoil. The second chapter is devoted to this subject.

The soil consists of arable surface soil and subsoil. In the former is accumulated the nutriment of plants chiefly cultivated for the food of man. This accumulation is affected by the absorptive power of the arable soil for mineral matters, by which soluble salts are removed from solution, and even chemical decomposition of the most stable compounds is brought about, and the bases or acids are retained by the soil in a firm state of combination. It is the presence of food in the soil in this state of *physical combination* which is alone available for the nutrition of plants. On the abundant or scanty supply of food in this state depends the fertility or sterility of a soil. In fertile soils food is present also in another form, in which it is not immediately available for the nutrition of plants. It exists as chemical compounds which are not soluble in water, or acids until rendered so by the action of powerful chemical agents, or to a much smaller extent by the slower process of the decomposing action of the weather. When the food is eliminated by disintegration (by fallow and mechanical operations) from this inert state of chemical combination, it passes into that of physical combination with the earthy particles before it is absorbed by the plant. Each kind of soil has its own absorptive power for causing the food to pass into a state of physical combination. When manure is applied, its greater or less dispersion throughout the soil will depend on this power. In general it is absorbed and fixed by the upper few inches of the soil, a smaller quantity penetrates to the lower layers, and scarcely any at all to the deep layers and subsoil. Hence when a subsoil is exhausted, manure cannot restore its fertility. From this peculiar property of soils of arresting the circulation of solutions of the food of plants, arises the necessity of employing means for the distribution of food, and for the uniform mixture of the different layers of the soil. The manner in which this is effected by mechanical operations, by organic matter, by manures, by certain chemical salts, &c., is pointed out in chapters second, third, and twelfth.

The quantity of food in a state of *physical combination* in any fertile soil is only limited. Continuous cultivation without replacement of all the mineral matters removed in the crops destroys fertility, either by causing the absolute loss of the assimilable food, or by altering the proper relative proportions between the different elements of food, to such an extent that the due growth of all parts of the plant is altered. For the successful growth of a plant in all its parts, every element of food is required. Not one substance has any superior fertilising power over another. The average crop of an unmanured field is always regulated by that element of food which is present in *minimum* quantity. The effect of a manure when beneficial is merely to increase the relative proportion of this *minimum* element. If the *minimum* matter was known in each case, its direct application would be sufficient to increase the fertility of the soil. But as in general this point is not ascertained, the application of farm-yard manure is certain in producing a fertilising effect, simply because it is a complex mixture containing *all* the food elements of plants, and consequently whilst supplying other matters which are not immediately wanted, it also furnishes the *minimum* substance. In chapter fourth, is discussed the question of this altered composition of the ground by cultivation.

In chapter eleventh, the fact that not one of the elements of food by itself possesses any superior nutritive value over the others is further discussed. Nitrogenous food, like all the rest, must be present if a plant is to grow properly, but no excess of this element of food will of itself produce more abundant crops. The analyses of soils show that they abound in nitrogen. Were all other sources of this element wanting, there would still be a continued supply provided for in rain and dew, and in the many processes of oxidation going on at the surface of the earth. Probably, wherever we have a generation and circulation of carbonic acid, there is also a provision for the formation of nitrogenous compounds. When Nature thus provides for a supply of nitrogen without the aid of man, it is likely that exhaustion of all other elements of food in the soil will take place by cultivation before this occurs with nitrogen. The inefficacy of the mass of nitrogen in the soil cannot be attributed to its existing in two forms, in one only of which it is assimilable. This is proved by experiments with soils and with farm-yard manure. When the nitrogen of the soil is not available, some other cause must be sought for than its existence in a state in which it is sparingly assimilable. This cause will be found to be the absence of some other elements of food, which, upon being supplied, will at once render the seemingly inoperative nitrogen at once energetic.

The diminution of the amount of available food elements in the arable surface soil, by the cultivation and sale of corn, necessitates the restoration of the removed mineral matters. This is effected to a limited extent by foreign manuring agents, but chiefly by the formation of manure by means of fodder plants. By the system of rotation, green crops which draw their nutriment from the subsoil are introduced between the cereals. By the deep penetrating roots of the former, the mineral matters of the subsoil are absorbed, and in the form of manure are transferred to the arable surface soil. But if this process continues, and the corn and cattle are still sold, and no replacement from without is made of the lost mineral matters, the time will arrive, sooner or later, when the subsoil becomes exhausted, and the surface soil having no longer a reservoir from which to draw supplies by means of fodder plants, is also unable to bear remunerative crops. This natural progress of the system of farm-yard manuring is fully discussed in chapter fifth. The reader must not suppose that the condemnation passed on the system of farm-yard manuring is meant to apply to farm-yard manure itself. The latter is the type of a valuable manure which cannot be replaced in every respect by any artificial mixtures in use. The remarks of the author only apply to the fallacious hopes entertained of keeping up permanently the fertility of the soil by manure obtained by the system of rotation, whilst we continue still to sell the corn raised by such manure without bringing back to the soil any portion of the mineral matter sold with the corn and cattle.

The excrements of man contain all the mineral matter not only of the corn, but also of the cattle sold from the land. Could we restore these excrements to the soil, a perfect circulation of the conditions of life for plants and animals would be established, and our fields would be retained in a permanent state of fertility. This problem has been solved by the Chinese and Japanese. Chinese rural life, as it is described by travellers, as well as the report of the Japanese system of husbandry given in Appendix G. by Dr. Maron, would scarcely lead us to wish for the improvement of agriculture upon the plan of these Orientals! The requirements of modern civilization would not permit the purchase of manuring matter, however valuable, at the cost of all domestic comfort. The sewers must, we fear, still receive what would be offensive to our English senses.

But can the contents of these sewers not be made available? The great mass of water which necessarily accompanies at present the fertilising matters, renders them of comparatively little value when compared with the expense of transport. But how to separate and concentrate these matters from the water is a problem which is at present occupying the earnest attention of scientific and practical men. The solutions hitherto proposed are far from satisfactory. The future of agriculture is, however, intimately connected with the right solution of this great sewage question.

In conclusion, I have only to state that the foreign weights and measures have been, when necessary, translated into their equivalents in English, but have been left unaltered when the point was only one of comparison, which could be equally illustrated by the foreign weights.

J. BLYTH, M. D.

QUEEN'S COLLEGE, CORK:
March 16, 1863.

PREFACE.

IN the sixteen years which have intervened between this work and the sixth edition of my 'Chemistry applied to Agriculture and Physiology,' I have had sufficient opportunity to become acquainted with all the obstacles which are opposed to the introduction of scientific teaching into the domain of practical agriculture. Among the chief of these may be reckoned the complete separation which has always existed between science and practice.

There has generally prevailed an idea that a smaller amount of information and intelligence is required for agricultural pursuits than for any other occupation; nay, that the practical skill of the farmer is only likely to be injured when he has recourse to science. Whatever requires thought and reflection is regarded as theory, which being the opposite of practice, must, of course, be of little value. The natural result of such opinions is, that when the practical man does attempt to apply scientific teaching, he is almost invariably a sufferer. He seems altogether to forget that man does not become intuitively acquainted with scientific teaching, which, like the skilful use of any complex instrument, must be learned.

The truth or error of the notions which guide our practice cannot, however, be regarded as a matter of indifference.

The more correct ideas which science has given us of the growth of plants, and the part played in the process by the soil, air, mechanical operations, and manure, is not regarded in the light of an improvement by the practical man, simply because his ignorance does not enable him to appreciate the information. Unable to find out the connection between scientific teaching and the phenomena presented in his daily pursuit, he naturally comes to the conclusion, from his point of view, that there really exists no connection between them.

The practical agriculturist is guided by facts observed in his own neighbourhood for a long period; or, if his views are more comprehensive, he follows certain authorities whose system of husbandry is held to be the best. It never enters into his thoughts to submit this system to proof, for he has no standard of comparison at hand. What Thaer discovered to be useful in Möglin was held to be equally so for all Germany, and the facts which Lawes found to be true on a very small piece of land at Rothamsted have become axioms for all England.

Under the dominion of tradition and of slavish submission to authority, the practical man has lost the faculty of forming a right conception of the facts which daily pass before his eyes, and in the end can no longer distinguish facts from opinions. Hence, when science rejects *his explanations of any particular facts*, it is asserted that *the facts are themselves denied.* If science declares that we have made progress in substituting for deficient farm-yard manure its active ingredients, or that superphosphate of lime is no special manure for turnips nor ammonia for corn, it is imagined that the utility of these substances is contested.

Long disputes have arisen about misconceptions of this

kind. The practical man does not understand the inferences of science, and considers himself bound to defend his own views. The contest is not about scientific principles, which he does not understand, but about the false conceptions he has formed of them.

Until this contest is ended by agriculturists themselves taking an active part in the matter, science can offer no effectual aid. I am doubtful if this time has yet arrived. I built my hopes, however, on the young generation who enter upon practice with a different preparation from their fathers. As for myself, I have reached the age when the elements of the mortal body betray a certain tendency to commence a new circle of action, when we begin to think about putting our house in order, and must defer to no later period what we have still to say.

As every investigation in agriculture requires a year before we shall have all the facts before us, I have scarcely any prospect of living to see the results of my teaching. The only thing that remains for me to do, under these circumstances, is to place my views in such a manner before the public, that there can be no possibility of misconception on the part of those who will give themselves the trouble of becoming thoroughly acquainted with them.

Many have reproached me with unjustly condemning modern agriculture as a system of exhaustion. From the communications addressed to me by many agriculturists as to their system of husbandry, I must exempt them from such an accusation. There are, however, but few among the general body who really know the true condition of their soil.

I have never yet met with an agriculturist who kept a ledger, as is done as a matter of course in other industrial pursuits, in which the debtor and creditor account of every acre of land is entered.

The opinions of practical men seem to be inherited like some inveterate disease. Each regards agriculture from his own narrow point of view, and forms his conclusions of the proceedings of others from what he does himself.

JUSTUS von LIEBIG.

MUNICH: *March*, 1863.

CONTENTS.

CHAPTER I.

THE PLANT.

CHAPTER II.

THE SOIL.

CHAPTER III.

ACTION OF SOIL ON FOOD OF PLANTS IN MANURE.

CHAPTER IV.

FARM-YARD MANURE.

CHAPTER V.

THE SYSTEM OF FARM-YARD MANURING.

CHAPTER VI.

GUANO.

CHAPTER VII.

POUDRETTE—HUMAN EXCREMENTS.

CHAPTER VIII.

EARTHY PHOSPHATES.

CHAPTER IX.

GROUND RAPE-CAKE.

CHAPTER X.

WOOD-ASH.

CHAPTER XI.

AMMONIA AND NITRIC ACID.

CHAPTER XII.

COMMON SALT, NITRATE OF SODA, SALTS OF AMMONIA, GYPSUM, LIME.

APPENDICES.

THE

NATURAL LAWS OF HUSBANDRY.

CHAPTER I.

THE PLANT.

Chemical and cosmic conditions of the life of plants—Conditions for the germination of the seed; moisture and oxygen, their action—Influence of the seed in the formation of the organs of absorption, and the production of varieties; influence of climate and soil in producing varieties—Importance of a knowledge of the developement of roots; radication of different plants—Comparison of the process of vegetation in annual, biennial and perennial plants—Growth of the asparagus, as an example of a perennial plant; storing of reserved food in its underground organs; use of this store—Meadow and woody plants—Growth of biennial plants; turnips: Anderson's experiments—Growth of annual plants; summer plants: tobacco; winter wheat, its developement like biennial plants; oats; Arendt's experiments; Knopp's experiments with maize in flower—The protoplastem (matter for forming cells); conditions for its formation; Boussingault's experiments; organic processes in plants, directed to the formation of the protoplastem—Absorption of food by plants not an osmotic process; marine-plants; duck-weed; land-plants; Hale's experiments on absorption by the roots and evaporation from the leaves—Power of the root to exclude certain substances from absorption not absolute; Forchhammer, Knopp—Comportment of the roots of land and water plants to solutions of salts; De Saussure, Schlossberger; comportment of land-plants to solutions of salts in the soil—Use of those mineral matters which are constant in different species of plants; iron, magnesia, iodine, and chlorine compounds—Absorption of matters by plants from the surrounding medium; influence of the consumption of them by the plant; part played by the roots in their absorption.

TO obtain a clear view of the theory and practice of Agriculture, we must keep in mind the most general chemical conditions of the life of plants.

Plants contain combustible and incombustible constituents. Of the latter, which compose the ash left by all parts of a plant on combustion, the most essential elements are—*phosphoric acid*, *sulphuric acid*, *silicic acid*, *potash*, *soda*, *lime*, *magnesia*, *iron*, and *chloride of sodium*.

The combustible constituents are derived from *carbonic acid*, *ammonia*, *sulphuric acid*, and *water*.

By the vital process of vegetation, the body of the plant is formed from these materials, which are therefore called the *food of plants*. All the materials constituting the food of our cultivated plants belong to the mineral kingdom. The *gaseous* elements are absorbed by the leaves, the *fixed* elements by the roots; the former, however, being often constituents of the soil also, may reach the plant by the roots, as well as by the leaves.

The gaseous elements form component parts of the atmosphere, and are, from their nature, in continual motion. The fixed elements are, in the case of land-plants, constituents of the soil, and cannot of themselves leave the spot in which they are found. The *cosmic* conditions of vegetable life are *heat* and *sunlight*.

By the cooperation of the *cosmic* and the *chemical* conditions, the perfect plant is developed from the germ or seed. The seed contains, within its own substance, the elements required to form the organs which are intended to take up food from the air and the soil. These elements are nitrogenous substances, similar in composition to the casein of milk or the albumen of the blood; and also starch, fat, gum, or sugar, with a certain quantity of earthy phosphates and alkaline salts. The farinaceous body, or so-called albumen of the seed of corn, as also the constituents of the cotyledons in leguminous plants, become the roots and leaves of the nascent plant. If corn-seeds are set to germinate in water, and allowed to grow upon a glass plate furnished with fine perforations, through which the roots may reach the water, the grain will go on growing for several weeks without receiving any incombustible element of food or any constituent of the soil. After three or four weeks the apex of the first leaf is seen to turn yellow; and upon examining the seed, nothing but an empty skin is found, for the starch has disappeared together with the cellulose (Mitscherlich). However, the plant does not die away, but new leaves are produced, often also a feeble stalk; the constituents of the first-formed, but now withering, leaves being applied to the formation of fresh shoots.

Under favourable circumstances, seeds with very large and vigorous cotyledons abounding in nutritive matter (e. g. beans) may, by vegetation in water alone, be got to flower—nay, even actually to produce small seeds; this developement, however, is mostly unattended by a perceptible increase of substance, but depends solely upon a mere transposition of the elements of the seed.

Nutrition is a process by which food is assimilated; a plant grows when its mass is augmented, and its mass is increased by absorbing materials from without, which are, from their nature, suited to become constituent elements of the body of the plant, and to sustain those functions upon which their assimilation depends.

The bud on a potato-tuber stands in the same relation to the constituents of the tuber as the germ in a corn-seed does to the farinaceous matter of the albumen. While the bud is developed in the formation of the young plant, the amylum and the nitrogenous and mineral constituents of the sap of the tuber are employed to form the young branches and leaves. A potato, which lay wrapt up in thick paper, in a box, in the Chemical Laboratory at Giessen—in a place absolutely dark, dry, and warm, where the atmosphere was seldom changed—was found to have produced, from each bud, a simple white shoot many feet long, showing no traces of leaves, but covered with hundreds of minute potatoes, which exhibited the same internal structure as tubers grown in a field; the cells consisted of cellulose, and were filled with minute starch granules. It is certain that the starch of the mother tuber, to have moved away from its position, must have become soluble; but it is equally clear that in the developement of the shoots a cause was operative within them, which (in the absence of all outward causes whereon growth depends) reconverted the dissolved constituents of the mother tuber into cellulose and starch granules.

The conditions required for the germination of a seed are—moisture, a certain degree of heat, and access of air; where one of these conditions is excluded, the seed will not germinate. By the influence of the moisture which the seed absorbs, and which causes it to swell, a chemical action takes place in it; one of the nitrogenous constituents acts upon the others, and upon the amylum, so that by a transposition of the elementary particles, the constituents are rendered soluble; the gluten is converted into vegetable albumen; the amylum and oil into sugar. If the oxygen of the air is excluded, the changes either do not take place or they proceed in a different way. The seeds of land-plants, when submersed under water, or placed in a soil covered with stagnant water, which excludes the air, will not put forth their plumules. This is the cause why many seeds, lying deep in the ground or in bogs, will remain for many years without germinating, although the conditions of moisture and temperature be favourable. It is often found that earth taken up from bogs, or brought up by the plough from the deep subsoil, and exposed to the atmosphere, becomes covered with vegetation, arising from seeds which, for their developement, required free access of air. Lowness of temperature tends to annul or retard the inflnence of the air upon the process of germination; whilst increase of temperature, with a proper supply of moisture, accelerates the chemical changes in the seed. No seed germinates below 32° Fahrenheit; each germinates at a definite temperature, and therefore in fixed seasons of the year. The seeds of *Vicia faba*, *Phaseolus vulgaris*, and the poppy, lose the power of germinating when dried at 95° Fahrenheit; while barley, maize, lentil, hemp, and lettuce seed retain the power at that heat; but wheat, rye, vetch, and cabbage seed will germinate even at 158° Fahrenheit.

During germination, oxygen is taken up from the air around the seed, and an equal volume of carbonic acid is evolved.

If seeds are set to germinate in glasses, with a slip

of litmus paper fastened on the inside, the paper is reddened, often after a very short time, owing to the disengagement of acetic acid: the most abundant and rapid evolution of free acid was found to take place in the germination of cruciferous plants, cabbage, and rape-seed (Becquerel, Edwards). Certain it is that the fluid contents of the cells of the roots, as well as the sap of most plants, have an acid reaction, from the presence of a non-volatile acid; the sap of the young spring shoots of the vine yields, upon evaporation, an abundant crystallization of bitartrate of potash.

By the experiments of Decandolle and Macaire, which have not yet been controverted, it was shown that vigorous plants of *Chondrilla muralis* and *Phaseolus vulgaris* which had been taken from the ground, with their roots, and were allowed to vegetate in water, imparted to the water, after a week's time, a yellowish tint, a smell like that of opium, and a harsh taste: whereas when the root was cut off at the stalk and both were placed in water, no such substances were given off as those which the entire plant had yielded.

Lettuces and other plants, when taken out of the gound, and, with their roots previously washed clean, are allowed to vegetate in blue litmus tincture, will continue to grow in the liquid, apparently at the expense of the constituents of the lower leaves, which wither away. After three or four days the litmus tincture assumes a red colour, which, however, disappears again upon boiling the fluid: this would seem to indicate that the roots had given off carbonic acid. If the plants are left longer in the litmus tincture, the latter suffers decomposition, and becomes neutral and colourless, while the colouring matter, separating in flakes, gathers round the fibres of the roots.

The developement of a plant depends upon its first radication, and the choice of proper seeds is therefore of the highest importance for the future plant. A crop of the same wheat, reaped in the same year, and from the same field, will exhibit differences in the size of the grains, some being larger, others smaller; and among both kinds, some when broken up will present a mealy, others a horny appearance, the one being more, the others less completely developed. The cause is this—that the stalks in the same field do not all shoot into ear and flower at the same time, and that some of them produce seeds much more maturely than others: hence the seeds of the one are far more developed, even in unfavourable weather, than the seeds of the others. A mixture of seeds unequal in their developement, or differing in the quantities of amylum, gluten, and inorganic matters which they severally contain, will produce a crop of plants as unequal in their developement as the original seeds from which they sprung.

The strength and number of the roots and leaves formed in the process of germination are (as regards the non-nitrogenous constituents) in direct proportion to the amount of amylum in the original seed. A seed poor in amylum will, indeed, germinate in the same fashion as another seed abounding in it; but by the time the former has succeeded, by the absorption of food from without, in producing roots and leaves as strong and numerous, the plant grown from the more amylaceous seed is again just as much more advanced in growth: its food-absorbing surface was larger from the beginning, and the growth of the young plant is in like proportion.

Poor and sickly seeds will produce stunted plants, which again will yield seeds bearing in a great measure the same character.

The horticulturist knows the natural relation which the condition of the seed bears to the production of a plant, which is to possess all or only some properties of the species: just as the cattle-breeder, who, with a view to propagation and increase of stock, selects only the healthiest and best-formed animals for his purpose; the gardener is aware that the flat and shining seeds in the pod of a stock gilly-flower will give tall plants with single flowers, while the shrivelled seeds will furnish low plants with double flowers throughout.

The influence of soil and climate gives rise to different varieties of plants, which, like races, are possessed of certain peculiarities, and are propagated by means of seed, as long as the conditions remain the same. Planted in another soil, or in a different climate, the new variety will lose again some one or other of its distinguishing characteristics.

The influence exerted by the condition of the soil in producing varieties of plants is observed most frequently with seeds that pass undigested through the intestinal canal of animals which have eaten them, and then receive a different manuring according to the various nature of the excrements of divers animals with which they are returned to the soil: an instance is afforded by the *Byrsonima verbascifolia* (v. Martius).

In the selection of seeds for planting it is always important to take into account the soil and climate from which they have been derived. In England seed-wheat from a poor soil is considered particularly well suited to a rich soil; rape-seed grown in colder regions or situations is sure to give a good crop in warmer localities. Clover seed and oats from mountainous districts are preferred to the same seeds from plains. Wheat from Odessa and from South Hungary is esteemed in colder regions also. The planters on the Upper Rhine import their hemp-seed from Bologna and Ferrara.

In like manner many German flax-growers, who wish to produce tall plants of uniform size, attach particular value to linseed from Courland and Livonia, where the soil and the nature of the climate, especially the short hot summer, bring the flowering and fruit time near together; so that the flowers, being simultaneously and uniformly fructified, produce ripe and perfect seeds.

Everyone knows how much the weather, during the flowering period, influences the formation of seed. If, after the flowering has commenced, cold weather or rain sets in, retarding the full developement of the inflorescence, the flowers fertilised at a later period produce no seeds, as the nutriment needed by them is applied by the flowers first fertilised for their own developement. It is a fact, that many plants will not repay the trouble of cultivation, if the climatic conditions are not sufficiently favourable to effect the thorough ripening of all the flowers, but serve only to ripen part of them.

With oats it often happens that in warm moist weather side-branches will spring from the axils of the leaves, when the principal culm is already shooting into ear; whence it happens, that at the end of the period of vegetation the plant is found to bear both ripe and unripe seeds.

The condition of the soil, as to porosity or compactness, influences the radication of plants. The fine filaments of the root, which are often coated with cork-like matter, are lengthened by the formation of new cells at their extremities, and they are obliged to exert a certain pressure, to force their way through the particles of earth.

The root-fibrils will always extend in that direction in which they encounter the least resistance; and this lengthening necessarily presupposes that the pressure wherewith the new-formed cells push aside the particles of earth, must be somewhat greater than the cohesion of the particles. The strength with which the root-fibres force their way through the soil, is not equally great in all plants. Those plants which have roots formed of very fine fibres are but imperfectly developed in stiff, heavy soils, wherein other plants with thicker and stiffer root-fibres will grow luxuriantly. The very resistance which the heavy soil opposes to the spreading of the roots of such plants tends to strengthen their fibres.

Of the cereals, wheat, with a comparatively feeble ramification of roots in the upper layers of the soil, still forms the strongest roots, which often penetrate several feet down into the subsoil; for a certain degree of compactness in the surface soil is favourable to the developement of its roots. There are instances on record, where parts of a wheat-field had been trampled down in the winter by horses (by no means an uncommon occurrence in the foxhunting districts of England), so far as to destroy every trace of a wheat-plant, and yet next year's crop turned out much more abundant on those very spots than in any other part of the field. It is evident that, to outlive an attack of this kind, a plant must have its principal roots spreading in the deeper layers of the soil. In the developement of its roots and the power of penetrating the deeper layers of the soil, the oat-plant stands next to wheat, and will flourish in a somewhat stiff soil; but as in the superficial layers also the roots of oats throw out a number of fine feeders, in a lateral direction, it is necessary that the top-soil should be rather light and open. A light, open loam, even if of no great depth, is particularly suited for barley, which forms a net-work of fine comparatively short root-fibres. Peas require a loose soil, with little cohesion about it, which will favour the spreading of the soft root-fibres in the deeper layers also; whereas the strong woody roots of the horse-bean will ramify in all directions, even in a heavy and more compact soil. Clover, grass-seeds, and small-sized seeds in general, put forth at first feeble roots of small extent, and require so much the greater care in preparing the soil, in order to ensure their healthy growth. The pressure of a layer of earth half to one inch thick suffices to prevent the developement of the seed sown in the ground. Such seeds require only just as much earth to cover them as will retain the needful moisture for germination. It is, therefore, found advantageous to sow clover together with corn of some kind; for as the corn is earlier and quicker in growth, its leaves shade the young clover plant, and protect it from the too intense action of the sun's rays; thus affording more time for the extension and developement of the roots. The nature of the roots * of rapes, turnips, and tuberous plants, clearly points out the part of the soil from which they draw their chief supply of food. Potatoes are formed in the

* Whenever the term 'root' is used in this work, the underground organs of plants are meant.

topmost layer of the soil; whereas the roots of beets and turnips, sending their ramifications deep into the subsoil, will succeed best in a loose soil of great depth. Still, they will also grow well in soil naturally heavy and compact, which has been properly prepared for their reception. Among turnips, the Swedish variety is distinguished by the numerous fibres which the root-stock sends into the ground; and mangelwurzel, with its strong and rather woody root-fibres, is still better suited than Swedes for a heavy clay soil.

On the length of roots but few observations have been made. In some cases it has been found that lucerne will grow roots thirty feet, rape above five feet, clover above six feet, lupine above seven feet in length.

A proper knowledge of the radication of plants is the groundwork of agriculture; all the operations which the farmer applies to his land must be adapted to the nature and conditions of the roots of the plants which he wishes to cultivate. On the root he should bestow his whole care; upon that which grows from it he can no longer exert any influence; therefore, to secure a favourable result to his labours, he should prepare the ground in a proper manner for the developement and action of the roots. The root is not merely the organ through which the growing plant takes up the incombustible elements of food required for its increase, but it may, in another not less important function, be compared to the flywheel in an engine, which gives regularity and uniformity to the working. It is in the root that the material is stored up to supply the growing plant with the needful elements for conducting the processes of life, according to the requirements made upon it by the action of light and heat.

All plants which give landscapes their peculiar character, and clothe the plains and mountain slopes with perennial green, have an underground developement, according to the geological or physical condition of the soil, admirably adapted to their perennial existence and propagation.

Whilst annuals are propagated and multiplied by

seeds alone, and have always a true root easily known by its simplicity of structure, by the absence of buds, and by the comparatively short range of its fibres, the turf and meadow plants are propagated by shoots and runners of a peculiar nature, and in many of them propagation is independent of the formation of seed.

As the strawberry, which will in a very short time cover extensive tracts of ground, sends forth from the stock above the root-bulb shoots in the shape of runners, which creeping along the ground, and producing here and there buds and roots, grow up as independent plants, so the perennial weeds, among which are here included the meadow and pasture plants, spread in a similar manner by corresponding underground organs. The creeping roots of the couch-grass (*Triticum repens*), the sea lyme-grass (*Elymus arenarius*), the trefoil (*Trifolium pratense*), the common toad-flax (*Linaria vulgaris*), propagate their plants by suckers in all directions from the mother-plant. The smooth-stalked meadow-grass (*Poa pratensis*) is propagated by a mother-stock, consisting of true roots, rooted runners, and creeping suckers; rye grass (*Lolium*) puts forth root-suckers in a stiff soil, and prostrate stolons in loose ground. Cat's-tail grass (*Phleum*) is found sometimes with bulbous, sometimes with fibrous many-headed roots, having a tendency to creep and to form mother-stocks. Timothy-grass grows stalk in the first year; in the second, it forms sometimes bulbous, sometimes fibrous many-headed mother-stocks, which send forth creepers in all directions. In the same way, meadow-grass spreads partly by budding suckers, partly by stolons.

On comparing the vital processes in annual, biennial, and perennial plants, we find that the organic work in perennials is principally directed to the formation of the root.

The seed of asparagus sown during autumn, in a fertile soil, will produce next year, from spring to the end of July, a plant about a foot high, the stem, twigs, and leaves of which from that time forward show no further increase. The tobacco plant, which is an annual, would from the same period to the end of August have produced a stem several feet high, covered with numerous broad leaves; and the turnip a broad crown of foliage.

But the cessation in the growth of the asparagus plant is only apparent; for from the moment that the external organs of nutrition are developed, the root increases in extent and substance in far greater proportion to the over-ground organs than is the case with the tobacco plant. The food which the leaves have absorbed from the air and the roots from the soil, having first been transformed into organisable matter, descends to the roots, in which there is gradually collected a sufficient store to enable the latter to furnish in the following year from themselves and without the least supply of food from the atmosphere the material required for the production of a new perfect plant, with a stem half as high again and a much greater number of twigs and leaves. The organic labour of this new plant, during the second year, results in the generation again of products which are deposited in the root, and, proportionately to the greater extent of the organs of nutrition, are stored up in much greater quantity than the roots had originally supplied.

The same process is repeated in the third and fourth years; in the fifth and sixth years the store deposited in the roots has become sufficiently rich to produce in spring, when the weather is warm, three, four, and more stems as thick as a finger, with numerous branches covered with leaves.

A comparative examination of the green asparagus plant, and of its withering stems in autumn, seems to indicate that at the end of the period of vegetation the remainder of the dissolved or soluble substances fit for future use, then still remaining in the overground organs, descend to the root. The green parts of the plant are comparatively rich in nitrogen, alkalis, and phosphates, whilst in the withered stems these substances are found in small quantities only. The seeds alone retain comparatively large proportions of phosphated earth and alkalis, being nothing else than the excess of those substances which the roots do not require for the next year.

The underground organs of perennial plants are the economic gatherers of all the vital conditions necessary for certain functions. If the soil will allow, they always collect more than they give out; they never spend all they receive. These plants form their flowers and seeds when the roots have collected a certain excess of phosphates, which may be given up without endangering the existence of the plant. An abundant supply of nourishment, by means of manuring, will accelerate the developement of the plant in one or another direction. Manuring a sward with ashes will draw from it clover plants; if acid phosphate of lime is employed, French rye-grass will spring up in thickly serried blades.

In all perennial plants, the underground organs are usually very much greater in mass and extent than those of annual plants. Whilst the roots of the latter die every year, the former preserve theirs in a state of readiness to absorb food at every favourable opportunity.

The circle from which a perennial plant draws its food enlarges from year to year; if one part of its roots finds little nourishment in a given spot, other parts draw their supply from other spots richer in the food required.

Only a very small portion of the plants of a thickly covered meadow will produce stems; the far greater part will develope only tufts of leaves; and many will for years be confined to the production of underground suckers.

For perennial grass and meadow plants, the production of underground suckers is of the highest importance, since by them the plant is furnished with nutriment at a time when a scarcity of supply would endanger the life of annual plants.

A good soil, and all other conditions of vegetable life, will of course exert the same favourable influence upon perennial as on annual plants; but the developement of the former is not so much dependent upon accidental and passing states of the weather, as is the case with the latter. Unfavourable conditions will, indeed, check the growth of a perennial plant, but only for a time, until a favourable change ensues, when the plant will resume growing; whereas an annual plant, under the same circumstances, reaches the limits of its existence and dies.

The permanence of vegetation on our meadows, and the certainty of their produce under varying conditions of soil and weather, must be attributed to the great number of plants which are able to continue for a shorter or longer period at a low stage of developement. While the one species of plants is developed above ground, producing flowers and seeds, a second and third species gather below the surface the conditions for a similar future growth. The one vegetation seems to disappear, to make room for another and a third, until for itself too the conditions for a perfect developement recur.

The woody plants grow and are developed in a manner quite similar to the asparagus plant, with this difference, however, that they do not lose their stem when the period of their vegetation comes to an end. An oak-sapling, 1½ foot high, was found to have a root above 3 feet long. The stem and the root serve jointly as a magazine for storing up the organisable matter to be used next year in restoring all the external organs of nutrition. When the stems of lime trees, alders, or willows have been cut down, they will, if lying in shady moist places, shoot out afresh, often after the lapse of years, and produce numerous twigs a foot long or more, covered with leaves.

The pauses which occur in the seed-bearing of forest trees are similar to those which are observed in most perennial plants, which, when growing on a poor soil, will also take several years to collect the conditions necessary for the production of fruit (Sendtner, Ratzeburg).

The loss of inorganic food-constituents, which the foliaceous trees suffer by the fall of the leaves, is trifling. When the leaves have attained their full formation, the cells of the bark receive a copious supply of amylum, which substance completely disappears from the cells in the boss of the leaf-stalk (H. Mohl). Even long before the fall of the leaves, their sap is considerably diminished, while the bark of the branches is, just at that time, often actually overflowing with sap (H. Mohl). In accordance with this fact, the analysis of the ash of the leaves shows that the amount of alkali and phosphoric acid in them decreases immediately before the fall; the fallen leaves contain such trifling quantities of these constituents, in comparison to their mass, that it is difficult to account for the injurious consequences arising from the raking up and removal of the fallen leaves in woods. (See Appendix A.)

A similar reflux of the assimilative products appears to take place in the grasses; when from the intense heat of summer the leaves begin to decay, chemical analysis reveals in the yellow leaves scarcely any traces of nitrogen, phosphates, and alkalis; and, indeed, animals instinctively turn from all kinds of fallen leaves, and refuse to feed on them.

In annuals and biennials the organic action results in the production of fruit and seed, after which the activity of the root comes to an end; in perennials, the production of seed is rather an accidental condition of their permanent existence.

The biennial can bestow more time than the annual in gathering the material necessary for the production of seed and fruit, which closes the period of its existence; but the time in which this takes place depends upon the state of the weather and the nature of the soil.

The annual is uniformly developed in all its parts; the food daily taken up is expended in increasing the overground and underground organs, which meanwhile take up a larger amount of food in proportion to the increase of their absorbent surface. With the growth of the plant, the conditions of increase inherent in the plant itself become enlarged, and exert their influence in proportion as the external conditions are favourable.

The developement of the biennial plants cultivated for their roots has three distinct periods; in the first period the leaves principally are formed; in the second, the roots, in which are stored the substances needed to produce the flower and fruit during the third period.

A series of experiments, made by Anderson, upon turnips, affords a clear view of the several directions in which the energy of a biennial plant tends at different periods of its growth. ('Journal of Agriculture and Transactions of the Highland Society,' No. 68, 69, new series, 5.)

These experiments were made to ascertain the total produce of vegetable substances obtained from turnips on one acre of ground. The turnips were gathered at four different stages of growth; the first on July 7, the second, on August 11, the third, on September 1, and the fourth, on October 5. The following table shows the weight of leaves and roots in pounds, taken up at the end of the respective stages, and calculated upon one acre of ground.

		Weight of leaves.	Weight of roots.
I.	Harvest after 32 days	219 pounds	7·2 pounds
II.	" 67 "	12,793 "	2,762 "
III.	" 87 "	19,200 "	14,400 "
IV.	" 122 "	11,208 "	36,792 "

The relative quantities of leaves and roots show that in the first half of the time of vegetation, sixty-seven days, the organic labour in the turnip plant is principally directed to the production and developement of the external organs.

From the 7th July to the 11th August, a period of thirty-five days, we find the increase to be 12,574 pounds in the leaves, and 2,755 pounds in the roots, which gives a daily increase of

Leaves.	Roots.
359 pounds.	78 pounds.

In this stage, accordingly, the production of leaves prevailed over that of roots to this extent, that out of eleven parts of food absorbed by the plants, nine parts went to the leaves and only two parts to the roots.

We find a very different proportion in the third stage; for during twenty days the weight of the leaves has increased by 6,507 pounds, that of the roots by 11,638 pounds, which gives a daily increase of

Leaves.	Roots.
325 pounds.	582 pounds.

During this third stage the plants take up daily somewhat more than double the amount of food taken up on any given day of the second stage, and this increase must stand in proportion to the daily enlargement of the surface of the roots and leaves; but the food absorbed is distributed in the plant in a very different manner. Of twenty-five parts by weight of food absorbed and assimilated, nine parts only remain in the leaves, the other sixteen parts serve to increase the mass of roots.

In exactly the same ratio as the leaves approached the limits of their developement, they lost the power of applying to their further growth the food which they had absorbed, and which now transformed into organisable matter was deposited in the roots. The same nutritive particles which went to form leaves, so long as the mass of foliage kept on increasing, now became constituent portions of the root.

This migration of the constituents of the leaves and transformation into constituents of the root appear to be most clearly shown in the fourth stage. The total weight of leaves, which on the 1st September still amounted to 19,200 pounds, had by the 5th October, or within the space of thirty-five days, decreased by 7,992 pounds, that is 228 pounds a day; in other words, out of every thirty-four leaves ten had withered, while the roots had increased by 22,392 pounds, or 640 pounds a day—a daily increase much more considerable than during the third stage.

It is evident that with the advance of autumn, with the lower temperature and diminished action of sunlight, the organic energy of the leaves decreased, and more than a third of the organisable matter collected in them descended to the roots, to be stored up for future use.

If we compare the quantities of nitrogen, phosphoric acid, potash, common salt, and sulphuric acid, absorbed during the last ninety days by the turnips growing on one acre of ground, we find from Anderson's experiments that the daily amount was as follows:—

Absorbed by the entire plant in a day.

Total increase.	Second stage.	Third stage.	Fourth stage.
			Pounds.
In substance	437	907	417
Nitrogen	1·15	0·695	1·21
Phosphoric acid	0·924	1·10	1·25
Potash	1·41	4·04	3·07
Sulphuric acid	1·12	1·57	1·52
Salt	0·84	1·98	1·11

Daily increase of roots in the fourth stage of growth.

	Phosphoric acid.	Potash.	Sulphuric acid.	Salt.
Supplied by the soil	1·25	3·07	1·52	1·10
" " leaves	0·41	1·56	0·51	0·53
	1·66	4·63	2·03	1·63

These figures show that the quantity of phosphoric acid taken up daily by the turnip plants growing on one acre of ground increases from the commencement of the second to the end of the fourth stage of growth, that is in ninety days from 0·924 to 1·25 pound a-day, which reckoned from one day to another makes the trifling difference of 0·0037 pound a-day.

Anderson suspects that his estimate of the nitrogen in the leaves during the third stage was not quite correct, and that it fell below the actual amount. If we add together the quantities of nitrogen absorbed in the last two stages, fifty-five days, we find a daily average of 1·02 pound of nitrogen, which is very nearly the same as in the preceding stage of growth.

The quantity of potash increased from the 11th August till the 1st September, in a somewhat higher ratio than the amount of vegetable substance produced. From the 1st September till the 5th October the increase of the roots was nearly double what it had been in the preceding stage, but this is explained by the migration of the potash compounds from the leaves to the roots. It is evident that the increase of potash has a certain connection with the formation of sugar and the other non-nitrogenous constituents of the roots, but no definite proportion can be established between them. The absorption of sulphuric acid increased uniformly in the three last stages; that of salt was a little greater in the third than in the second and fourth stages.

Without wishing to indicate the exact part performed in the process of vegetation by these various mineral substances, as also by lime, magnesia, and iron, we remark that, except in the case of potash, the absorption of them was evidently uniform from day to day, yet showing every day a trifling increase corresponding to the daily increase of the food-absorbent surface up to the fourth stage of growth.

The smallest increase was seen in phosphoric acid and nitrogen, both equally necessary for the formative processes going on in the turnip plant; and it is manifest that they must have served to bring into operation some more powerful agency, whose effects are revealed in the production and augmentation of the non-nitrogenous constituents.

If we take the quantity of mineral substances absorbed as the measure of their importance for the organic operations going forward in the plant, we must assign to sulphuric acid and common salt an influence equal to that of any of the others.

Looking at the qualities of mineral constituents severally taken up by the different parts of the plant in the various stages of growth, we observe the greatest disparities. In the second stage, a quantity of potash, amounting in the aggregate to 49·29 pounds, was absorbed in 35 days; and of this, the roots were found to contain 8·02 pounds, equal to one-sixth—the leaves 41·27 pounds, equal to five-sixths. The same proportion—namely, about five to one—was found to exist between the weight of the leaves produced, and that of the roots.

In the third stage, the weight of the roots produced exceeded that of the leaves; and of the 80 pounds of potash absorbed by the plants, 34 pounds, or more than one-third, remained in the roots. The same was found to be the case with phosphoric acid, and the other mineral constituents; that is to say, they were found distributed in varying proportions, corresponding to the growth and increase of the mass of the overground and underground organs of the turnip plants, which, in the various stages, are likewise not uniform.

If we regard the mere increase of the leaves and roots in mineral substances, without reference to the total amount of them absorbed by the entire plant, it appears to be most irregular, and to proceed by 'fits and starts.' The plant receives every day nearly the same quantity of phosphoric acid, nitrogen, salt, and sulphuric acid, which are distributed in the several parts of the plant, leaves, or roots, where they are required for use. The chief difference observable is in the increase of potash, which in the third stage is out of all proportion greater than that of the other mineral constituents.

It is highly probable that from the raw material—i. e. the carbonic acid, water, ammonia, phosphoric acid, sulphuric acid, with the cooperation of the alkalis, earths, &c.—the chemical process engenders in the plant simply a nitrogenous and sulphureous substance, belonging to the albumen group, and only one non-nitrogenous substance, belonging to the group of hydrocarbons. The former retains its character during the period of vegetation; while the non-nitrogenous substance is converted into a tasteless, gum-like body, or into cellulose, or sugar—becoming a constituent of the leaves or of the roots, according as the organic energy preponderates in the overground or underground organs.

If there is a relation between the phosphoric acid and the production of the nitrogenous constituents, the soil must contain, in its parts, definite proportions of both substances; and for the cultivation of turnips, the upper layers must necessarily be much richer in phosphates than the lower. For in the first half of the period for vegetation, the branching of the roots is much less extensive than at a later period, and the root is in contact with a much smaller bulk of earth than afterwards; hence, if the root is to draw from this smaller bulk the same amount of nourishment as from the larger, the former must contain more of it, in proportion as the absorbent root-surface is smaller.

The ash of all plants in whose organism large quantities of amylum, gum, and sugar are produced, is distinguished from the ash of other plants by the preponderance of potash; now, if the potash in the sap of the turnip plant formed a necessary agent in the formation of sugar and the other non-nitrogenous constituents, the quantity of that mineral matter absorbed in the third and fourth stages of growth is easily explained—because the formation of the non-nitrogenous constituents of the root was more active in these than in the former stages.

That the production of the combustible constituents—the conversion of the carbonic acid and ammonia into non-nitrogenous and azotised substances—stands in a definite relation of dependence to the incombustible matter found in the ash, is an opinion which no longer requires special proof to support it. But the dependence is mutual. To say that the reason why the azotised or non-nitrogenous products are formed in large proportion is *because* the plant has taken up more phosphoric acid or potash, is just as correct as to assert that the plant takes up more phosphoric acid or potash *because* the other conditions required for the production of azotised or non-nitrogenous substances are found combined in its organism.

To enable a plant to attain its maximum of growth, the soil must at all times yield, in an available form, the whole quantity of each of its constituents; and, on the other hand, the cosmic conditions—heat, moisture, and sunlight—must cooperate to transmute the absorbed substances into the organs of the plant. If the substances that have passed from the soil into the plant cannot be turned to account, from the want of this cooperation, no fresh substances are absorbed; in unfavourable weather, the plant does not grow. No more does it grow, even though the outward conditions are favourable, if the soil contains no proper nourishment.

In the second half of the period of developement, the roots of the turnip plant, having penetrated through the arable surface deep into the subsoil, absorb more potash than in the preceding stage. If we suppose that the absorbing spongioles of the root reach a stratum of soil poorer in potash than the upper layer, or not sufficiently rich in that material to yield a daily supply commensurate with the requirements of the plant, at first, indeed, the plant may appear to grow luxuriantly; yet the prospect of an abundant crop will be small, if the supply of the raw material is constantly decreasing, instead of enlarging with the increased size of the organs.

In the economy of the turnip, the root receives during the last month of vegetation nearly one-half of all the movable constituents of the leaves; and this constitutes, after the completion of its first year's period of vegetation, a store of organisable matter for future use.

During the spring of the following year the root begins to shoot, putting forth a slight leafy top, and a flower-stalk several feet high; with the developement and maturing of the seed, the plant dies. The chief bulk of the food stored up in the root is applied, in the second year or third period, in quite a different direction; though, beyond the mere supply of water, the soil seems to take no part in this new act of life.

All monocarpous plants—that is, all plants which

flower and produce seed but once—present, like the turnip plant, distinct periods of life, as regards the direction of organic activity in them. In the first, the plant produces the organisable matter required in the succeeding period; in the latter, that which is required for the final functions of life. But these materials are not always stored up in the root, as is the case in the turnip; in the sago-palm they fill the stem; in the aloe (*Agave*) they collect in the thick fleshy leaves.

The production of seed is, with many of these plants, much less dependent upon any fixed period of time, than upon the store of organisable matter collected in them in the time preceding. Favourable climatic conditions or propitious weather will hasten, while unfavourable cosmic conditions will retard, its production.

The so-called summer-plants are monocarps which are able to gather in a few months the conditions required for the production of seed. The oat-plant grows to maturity and bears ripe seed in ninety days; the turnip-rape only in the second year of its existence; the sago-palm in sixteen to eighteen years; the aloe in thirty to forty, often not till 100 years. (See Appendix B.)

In many perennial plants, the outer part dies every year, while the root lives on. In the monocarpous plants, the root dies with the production of the seed. In these the production of seed is an *indispensable*, in the perennial plants more of an *accidental*, condition of continued existence.

The economy of plants is regulated by laws which manifest their operation in the peculiar faculty of certain organs to store up food for future use; so that all the external causes which seem to hinder their developement, actually contribute in the end to insure their continued existence, *i. e.* their propagation.

The contents of the roots in perennial grasses and asparagus, may, in the different periods of the life of these plants, be compared to the farinaceous body or albumen in the grain of cereals; with this difference, however, that the skin does not become empty as is the case with the latter on germination, but is always refilled and keeps increasing in size. The perennial plant always receives more than it expends; whereas the monocarpous plant spends its whole store in forming fruit.

The fact that the roots of the turnip, in autumn, grow at the cost of the constituents of the leaves, readily explains the influence which the removal of leaves will exercise upon the crop at different stages of growth. The removal of a few leaves in August makes no great difference to the root, while the removal of leaves at the end of September causes the greatest damage to the root-crop. Metzler, who made very accurate comparative experiments upon this point, found that an early cutting of the leaves reduced the turnip crop by 7 per cent. only, while a late, or a second cutting, reduced it by as much as 36 per cent.

If, in the first year, instead of the turnips being removed from the field at harvest, the tops were merely cut off and the roots were left and ploughed in, the field would, on the whole, sustain a loss of soil constituents; still the roots in the soil would retain the greater portion of them. A very different relation would arise, if at the end of the second year of vegetation the turnip tops were cut off, and the stem were removed together with the seed. For, at the end of the first year, the root would still retain the far larger portion of the azotised and also of the incombustible constituents, which would thus be left in the soil; but in the second year these materials would be carried into the overground part of the plant, and there be used for the production of the stem and the seed; hence, the removal of the latter would of course make the soil poorer, even though the roots were now left in it. Before the shooting and flowering, the root was rich in soil constituents; after the production of seed, its store of them is exhausted. If the plant is cut off and the root left in the ground, before flowering, the soil retains the far greater portion of the nutritive matter which it had given to the plant; on the contrary, after flowering and the production of seed, the root retains only a small residue of these constituents, and the soil is correspondingly exhausted of them.

As it is with the turnip, so is it with culmiferous plants. If they are cut off before flowering, a considerable portion of the nutritive substances stored up in them remains in the root, which the soil of course loses, if the overground plant is removed after the ripening of the seed.

The experience derived from the cultivation of tobacco gives a clear view of the processes in the developement of an annual leafy plant.

In the tobacco plant the overground and the underground parts grow with perfect equality; the root gains in extent, in the same proportion as the stem lengthens and the leaves increase in number and size. There is no appearance of sudden changes in the direction of organic activity, no shooting, but the phases of life in the plant follow in steady continuous progression. Even while the top of the stem bears ripe seeds, and the lower leaves have withered, the side shoots of the plant are often still putting forth flower-buds, the seeds of which will ripen at a much later period.

The tobacco plant is remarkable for producing in its organism two nitrogenous compounds, of which the one, nicotine, contains neither sulphur nor oxygen; while the other, albumen, is identical with the sulphureous and oxygenised constituents of the cereals and other alimentary plants.

The commercial value of tobacco leaves is in an inverse ratio to the amount of albumen which they contain, that sort of tobacco being most highly esteemed by smokers which contains the least albumen; for the latter ingredient, in the burning of the dry leaves, emits on carbonisation a most disagreeable smell of burnt horn shavings. The leaves rich in albumen contain, as a rule, more nicotine than those which are poor in albumen; they give the strongest kinds of tobacco, many of which cannot be smoked unmixed.

The tobacco leaves cultivated in France and Germany are manufactured either into smoking tobacco or into snuff. For the fabrication of snuff, leaves which are rich in albumen and nicotine are preferred to those containing a smaller amount of those ingredients. The leaves intended for snuff are, either when still entire or after being ground to powder, subjected to a kind of fermentation, which takes place pretty speedily, with evolution of heat, if they are kept moistened with water. From the putrefaction of the albumen there arises a considerable quantity of ammonia, which is a principal ingredient of German snuff, and is also occasionally increased by the manufacturers, by moistening with carbonate of ammonia or caustic ammonia, to suit the taste of consumers.

The leaves intended for smoking are also improved in quality by a slight process of fermentation, which serves to diminish the quantity of albumen in them.

These preliminary remarks will help to explain the different methods of cultivating tobacco.

The size of the leaf in length and breadth, its light or dark colour, the height of the stem, the amount of produce, and the greater or less proportion of albumen and nicotine, all depend very essentially upon the manuring of the plant.

The plant succeeds best, in Europe, on light, sandy, humose, loamy, or marly soils. The strongest kinds, richest in albumen and nicotine, are grown on virgin land, and on heavy clay soil manured with bone-dust, shavings and clippings of horns and claws, blood, bristles, human excrements, oilcake, and liquid manure.

In Havannah, tobacco is grown on virgin soil, on cleared forest lands, which are often burnt first, as is done in Virginia. The best qualities (the poorest in albumen) are yielded in the third year of cultivation.

From this it would appear, that animal manure abounding in nitrogen (ammonia) favours the production of nitrogenous constituents; but the soil, on the other hand, which is poor in ammonia, and probably contains the nitrogen in the form of nitric acid, produces leaves containing much less albumen and nicotine.

The effect of removing the tobacco plant from the rearing beds to the field is very striking. Transplanted into the new soil, the young tobacco plant proceeds in the first instance, like seed in the process of germination, to produce roots; the leaves already formed wither on transplantation, and their movable constituents, together with the store of organisable matter collected in the roots, are applied to the production of numerous branch radicles. A second transplantation has a still more favourable effect upon the underground organs of absorption.

As the direction of the organic operations in summer-plants is entirely turned to the formation of seed, and as this consumes the materials which give activity to the roots and leaves, the tobacco planter breaks out, when the plant has put forth six to ten leaves, the heart of the middle stem, on which the flowers and seed capsules grow. Stripped thus of the crown, the whole vigour of the plant is now directed to the buds between the leaves and stem, and these put forth side-shoots which are treated like the principal stem, that is to say, they are either broken away, or simply cracked by twisting. Thus the leaves retain the organisable matter subsequently produced, and increase in mass and size, while the amount of water in them diminishes. By the middle of September, the leaves lose their green colour and are spotted with yellow blotches, imparting a marbled look; they become parchment-like, feel dry to the touch, get flaccid, with the ends drooping to the ground, and, when arrived at full maturity, are viscous, clammy, and readily come off the stem.

This treatment is variously modified, according to the several varieties of tobacco, and the different countries in which it is grown. The so-called common English tobacco, which is particularly rich in nicotine, is often allowed by planters to run to seed, in order to effect a separation of the nitrogenous constituents, the albumen forsaking the leaves and lodging in the seed.

In the young shoots, buds, and generally in all parts in which the production of cells is most actively carried on, the sulphureous and nitrogenous constituents (albumen) accumulate, and thus the younger leaves are always richer in these substances than the older. The leaves nearest the ground (sand-leaves) give a milder, the upper leaves a stronger tobacco. In those varieties which are not particularly rich in nicotine and albumen, the sand-leaves are of much less value than the upper leaves. A mild tobacco always means a tobacco poor in narcotic constituents.

The course pursued by the European tobacco planter, who lays a superabundance of animal manure upon his fields, is the exact reverse of that adopted by the American planter, who cultivates his plants upon a field that has never been manured. The one seeks to reduce or dilute the narcotic, sulphureous, and nitrogenous constituents of the leaves; the other to concentrate them. Accordingly, the American planter breaks the lower leaves in their full vigour, when the plant has attained to half-growth; the European planter attaches the greatest value to the fully-developed upper leaves.

As the tobacco plant, like all annuals, only yields up its whole store of organisable matter at the ripening of the seeds, the stem does not die after the loss of the leaves; but the materials still remaining in it and in the roots cause the stem to send forth fresh shoots, and frequently even leaves, though small-sized ones. In the West Indies, Maryland, and Virginia, before the gathering of the leaves, the stems are notched immediately above the ground, so that they lean over without being severed from the root. In warm weather, the water in the leaves evaporates, and a motion of the sap ensues from the stems and roots towards the leaves, in which the sap is thus concentrated as the plant withers. The tobacco planters on the Rhine have found that a superior tobacco, poorer in albumen and nicotine, is produced if, instead of breaking the leaves off in the field, the plant with the leaves on it is cut down just above the ground, and hung up to dry with the top downwards. The stem will, under these circumstances, continue to vegetate for a time, sending forth small shoots which gradually turn in an upward direction and put forth flower-buds. In these flower-buds the sulphureous and nitrogenous constituents are collected from the leaves, which lose these ingredients in the same proportion, and are thereby improved in quality.

Of the plants cultivated for the sake of their seed, wheat holds the chief place.

Winter-wheat is in its developement extremely like a biennial plant. In the biennial turnip we see that with the first leaves a corresponding number of root-fibres are produced; and that after the formation of the leaf-top, the root begins to expand greatly in size and extent, immediately after which the flower and seed-stalk shoots forth.

Very soon after winter-wheat is sown, the young plant puts forth the first leaves, which in the course of winter and the early months of spring increase to a tuft; to all appearance the vegetation of the plant seems to cease for weeks and months. When warm weather comes, the plant puts forth a soft stem, several feet high, furnished with leaves, and bearing at the top an ear set with flower-buds in which, after flowering, the seeds are formed. As the seed is developed, the leaves from the bottom upwards turn yellow, and die with the stem as the seed ripens.

It cannot be doubted that while the growth of the plant appears to have ceased before the time of shooting, the over and under ground organs are in constant activity; food is incessantly absorbed, which, however, is but partially employed to increase the mass of leaves, but not to form the stem. There is, therefore, every reason to believe that the far larger portion of the organisable matter produced in the leaves during this period goes to the roots, and that this store is afterwards applied to the formation of the stalk. On the approach of warmer weather all the operations of life in cereal plants are quickened, and the quantity of food daily absorbed and worked up increases with the extent of the absorbing and elaborating organs. In spring many of the older leaves and of the root-fibres die in the portions of the soil exhausted by them; the root-tops send forth new buds, and with every new bud new rootlets, until the stalk-joints have attained a certain length. From this time forward to the end of the period of vegetation, both the food absorbed by the plant, and the movable part of the materials formed in the leaves, stem, and root, go to form flowers and seeds.

The observations of Schubart show that the roots of cereal plants, in the first period of vegetation, increase much more than the leaves. Schubart found that rye plants, which, six weeks after sowing, presented leaves 5 inches long, had meanwhile produced roots 2 feet in length.

The vigour with which cereal plants send forth their stalks and side-shoots corresponds to the developement of the root. Schubart found as many as eleven side-shoots in rye plants, with roots 3 to 4 feet long; in others, where the roots measured $1\frac{3}{4}$ to $2\frac{1}{4}$ feet, he found only one or two; and in some, where the roots were but $1\frac{1}{2}$ foot, no side-shoots at all.

The action of a low temperature in autumn and winter, which puts a certain limit to the activity of the outer organs, without altogether suppressing it, is essential to the vigorous thriving of winter corn. It is a most favourable condition for future developement, if the temperature of the air is below that of the soil, so as to retard for several months the developement of the outer plant.

Hence a very mild autumn or winter operates unfavourably upon the future crop, as the higher temperature encourages the developement of the principal stalk before the proper time, which shoots up thin, and consumes the food which should have served to form buds and new roots, or to increase the store of organisable matter in the roots. Thus stunted in its developement, the root supplies less food to the plant in spring, as it takes up and gives out less in proportion to its smaller absorbent surface and more limited supply stored up in it; and it retains the same feeble character in the succeeding periods of vegetation. The agriculturist endeavours to meet the difficulty by grazing down or cutting these feeble plants; the formation of buds and roots hereupon begins anew, and if the external conditions are favourable, and the plant has time to fill the root with a fresh store of organisable matter, the normal conditions of growth are, in the agricultural sense, restored. Summer corn maintains, in the several periods of its developement, the same character as winter corn; only these periods are of much shorter duration.

Ahrend's study of the oat-plant in its several stages of growth is instructive in this respect. He determined the increase in combustible and incombustible constituents during the following periods: from germination to the beginning of shooting (end of the first stage, 18th June); from this time to shortly before the end of shooting (second stage, 30th June); immediately after flowering (third stage, 10th July); the commencement of ripening (fourth stage, 21st July); finally, to perfect maturity (fifth stage, 31st July). On the 18th June the plants were on an average 31 centimeters high (1¼ inch), the three lower leaves were nearly expanded, the two upper leaves were still folded up. Of the stalk-joints the three lower alone had an appreciable length (1, 2, and 3 centimeters), the three upper had but a rudimentary existence. Twelve days after (on the 30th June) the plant had attained double the height (63 centimeters); and ten days after this again, on the 10th July, after flowering, it had reached 84 centimeters.

1,000 plants respectively produce in grammes:—

Constituents	Examined on				
	18th June. I. stage. In 49 days, before shooting.	30th June. II. stage. In 12 days, stalks full grown.	10th July. III. stage. In 10 days, flowering.	21st July. IV. stage. In 11 days, formation of seed.	81st July V. stage. In 11 days, ripening.
	Grammes.	Grammes.	Grammes.	Grammes.	Grammes.
Combustible...	419	873	475	435	128
Incombustible..	36·6	33·48	30·33	20·34	7·18
	In one day.				
Combustible...	8·551	72·75	47·50	39·45	12·8
Proportion ..	1	8·5	5·5	4·6	1·5
Incombustible..	0·747	2·79	3·03	1·849	0·318
Proportion ..	1	3·73	4·06	2·47	0·96

In looking at these figures we must remember that Ahrens could only determine what the overground part of the plant had received from the root, not, as Anderson in the case of the turnip, what the whole plant had derived from the soil. The great disparity in the increase of combustible and incombustible substances evidently depends rather upon the unequal distribution of the materials absorbed, than upon any disparity in the quantity derived from the soil. The whole period of developement comprised about 92 days, and we see that for more than the first half (49 days) the plant remains stationary at an apparently low stage of growth, the foliage alone being developed, and that not fully. In the next 12 days, from the 18th to the 30th June, the plant gains double the weight of incombustible constituents, and grows twice as high as in the 49 days preceding; and within this short time, the overground parts absorb nearly the same quantity of incombustible constituents as they had previously taken up. In fact, the plant takes up 8½ times the quantity of combustible matter, and 3¾ times more of ash constituents on one day of shooting, than upon one of the 49 previous days.

We cannot suppose it at all likely that the external conditions of nutrition, the supply of food by the atmosphere and from the ground, or the absorptive power of the plant, should alter and increase, by fits and starts, from one day to another. We are led rather to assume that the oat-plant is subject in its developement to the same law which we have observed in the case of the turnip, and that therefore, in the second half of the first stage of growth, the activity of the leaves was principally directed to the production of organisable matter, to be stored up in the root in the shooting stage, and then supplied to the overground organs of the plant. The heightened assimilative or working power of the plant, consequent upon the higher temperature and brighter sunshine of summer, was attended by a proportionate increase in the supply of food; but the relative proportion of the soil constituents remained much the same as in the turnip plant.

If we compare the respective quantities of potash, phosphoric acid, and nitrogen, which the overground parts of the oat-plant have received from the root and the soil, in the several stages of growth, i. e. to the commencement of flowering, thence to incipient ripening, and finally to maturity, we find that 1,000 plants have received:—

	In the I. and II. stages, 61 days.	In the I. and II. stages, 21 days.	In the V. stage, 10 days.
	Grammes.	Grammes.	Grammes.
Potash	34·11	13·2	0·0
Nitrogen	25·0	24·9	5·4
Phosphoric acid	5·99	6·94	1·33

These proportions show that the daily increase of potash in the overground parts of the oat-plant was pretty nearly the same in the 21 days of the 3rd and 4th stages, as in the 61 days of the 1st and 2nd. But for the phosphoric acid and the nitrogen a very different result is obtained; we find that the quantity of these two ingredients which passed into the stalk, the ear, and the leaves, amounted in the 21 days of the 3rd and 4th stages to as much as in the 61 days of the 1st and 2nd stages: in other words, the overground organs of the plant gained of these two ingredients, in the flowering and ripening time, three times as much each day as in the preceding period.

Of the turnip-plant we know with tolerable certainty, that from the time when it sends forth a flower-stalk, the constituents of the stalk, as also those of the flowers and the seeds, are for the most part stored up in the root, and are supplied therefrom. It is highly probable that the corn-plant is similarly circumstanced, and that from the flowering to the end of life it is fed, though not exclusively, by the root, which from the flowering time gives out what it had stored up in the preceding period.

Knop observed that Indian corn plants in flower, taken out of the ground and placed with their roots simply in water, produced ears with ripe seeds; which proves that the materials serving for the production of seed were already present in the plant at the time of flowering.

It is an established fact that a corn-plant, if cut off before flowering, relapses into that lower stage of vegetation of a perennial plant, in which the root receives more organisable matter than it parts with.*

The proportions of incombustible constituents and of nitrogen severally required by oats and turnips, are remarkably different both in the aggregate and during the various stages of growth. The facts established by Anderson for the turnip, and by Ahrends for the oat, are indeed not sufficiently numerous to warrant us in deducing any positive law of growth for those two plants: still a few inferences may easily be drawn from them. The quantities of phosphoric acid and nitrogen in the turnip are, at the end of the first year of vegetation, nearly in the proportion of 1 : 1; in oats, on the contrary, of 1 : 4. The oat-plant requires to the same quantity of phosphoric acid four times as much nitrogen

* Buckmann ('Journ. of the Royal Agric. Soc.') sowed wheat on a field in autumn 1849, which was continually cut down in 1850, so that the plants were never allowed to come to flower: they were left in during the winter 1850–51, and yielded an excellent crop in the year 1851.

as the turnip; and the latter to the same quantity of nitrogen four times as much phosphoric acid.

If the developement of the oat-plant takes a similar course to that of the turnip, the former must have accumulated in its underground organs before the time of shooting a store of organisable matter, similar to that laid up by the turnip at the close of the first year of vegetation. The mass of organic substances accumulating in these plants before the developement of the flower-stalk is manifestly much larger in the turnip than in the oat-plant. The former receives from the soil much more phosphoric acid and nitrogen; but the turnip had 122 days, the oat-plant only 50 days, up to the period of shooting for extracting these nutritive substances from the ground. Now if the turnips and oats growing on a hectare (2½ acres) of land had daily received an equal amount of them, then, all other circumstances being the same, the quantity of nutritive substances absorbed would be proportionate to the time of absorption. In this respect the nature of the root makes a great difference, according to the extent of absorbent root-surface. The larger root-surface is in contact with more earthy particles, and can during the same time extract more nutritive substances than the smaller. The mass of vegetable substance produced, and especially the quantity of non-nitrogenous and azotised materials, depend upon the nature of the plants. If the absorbent root-surface of the oat-plant were 2·45 times greater than that of the turnip, the former would, under like circumstances, take up daily 2·45 times as much food as the latter, i. e. the oat-plant would absorb in 50 days as much as the turnip in 122 days. Thus in equal times the power of two plants to absorb food is in proportion to their absorbent root-surface.

The time of vegetation occupied by the turnip-plant comprises, in the first year, 120 to 122 days, and terminates at the end of July in the next year with the production of seed. If we take the whole time of vegetation of the turnip plant at 244 days, and suppose the time of vegetation of the oat-plant extended from 93 or

95 to 244 days, we find that this would give sufficient time for growing two oat crops, and advancing a third half way to maturity; and a careful investigation might perhaps reveal that the quantity of sulphureous and nitrogenous constituents produced in the oat-plant is not less than that obtained in turnip-plants from an equal area of ground.

In the grains of the cereals, the quantity of the sulphureous and nitrogenous constituents is to that of the non-nitrogenous (the quantity of the blood-making substances to the amylum), as 1 : 4 or 5; in the roots of turnips, or in the tubers of potatoes, as 1 : 8 or 10. In the latter, therefore, the quantity of the non-nitrogenous constituents is in proportion to the other constituents much greater.

When at a certain temperature the organic process of germination begins in a grain of wheat, the embryo first sends down a number of rootlets, while the plumule rises upward in the form of a short stem, with two or three complete leaves. Simultaneously with the changes taking place in the embryo, the constituents of the farinaceous body (albumen) become fluid; the amylum is converted first into a substance resembling gum, then into sugar, while the gluten changes into albumen, and both together form protoplastem (Naegeli's organic food elements), or the food of the cell. The fluidity of the new body enables it to find its way to the places where the formation of cells is going on. The amylum supplies the elements required to form the outer wall of the cell; the nitrogenous matter constitutes a principal ingredient of the cell contents. Simultaneously with the roots and leaves, small leaf-buds arise upwards on the stem-joint, and small root-buds appear at the basis of the roots.

In the protoplastem of the wheat-plant the non-nitrogenous matter exceeds the azotised matter as five to one.

Except water and oxygen, no substance from without takes any part in these processes. What the seed loses in carbon by the formation of carbonic acid in germination is afterwards recovered almost entirely by the young plant.

The plant developed under these circumstances barely increases in substance to any appreciable degree, even though it may continue vegetating for weeks. The organs developed from a grain of wheat weigh all together, when dried, no more than the grain did before germination. The relative proportion of the non-nitrogenous and azotised substances in them is almost the same as in the farinaceous body, the constituents of which have in reality merely assumed other forms. The leaves, roots, stem, leaf-buds and root-buds collectively represent the constituent parts of the seed, transformed into organs and apparatus now endued with the power of performing certain operations which serve to carry on a chemical process, whereby external inorganic substances, with the cooperation of sunlight, are converted into products analogous in all their properties to the materials from which these organs themselves arose.

The organic process of cell-formation presupposes the presence of the protoplasm, and is independent of the chemical process by which the latter is generated; but this chemical process is indispensable to the continuance of the cell-formation.

In a young plant which has been developed in pure water alone, the chemical process must soon come to an end for want of the necessary external conditions. The leaves and roots in this case can do no work as formative organs. In the absence of food they generate no products upon which the continued existence of the plant depends. When they have arrived at a certain state of developement, the cell-formation ceases in themselves, although it is still continued in the new root-buds and leaf-buds. The latter stand to the movable contents of the previously existing leaves and roots in the same relation as the embryo of the wheat-seed to the farinaceous body. The non-nitrogenous and azotised constituents which represent the working capital of the existing roots and leaves are transformed as these die into new organs, and new leaves are developed at the expense of the constituents of the old ones. But these processes are of short duration; after a certain number of days the young plant dies. The more immediate external cause of its short duration is the want of food; but another *internal* cause is the conversion of the non-nitrogenous soluble substances into cellulose or woody tissue, whereby it loses mobility. With the diminution of this soluble substance the most essential condition of cell-formation is impaired: when the whole has been consumed, the process comes to an end. The withered leaves, when burnt, leave behind a certain quantity of ash, showing that they retain some mineral matter; there remains in them also a small portion of nitrogenous substance.

The most remarkable thing in this developement is the part performed by the nitrogenous matter of the seed, which becomes a constituent element of the root-fibres, stems, and leaves, where its agency serves to bring about the formation of cells. After the death of the first leaves, it becomes a constituent of the new ones, performing in them the same part over again, so long as there remains materials for cell-formation. But the nitrogenous matter itself is not in reality worked up in the plant, and forms no actual tissue or component part of the cell.

The experiments of Boussingault on the growth of plants, in the absence of all nitrogenous food ('Annal. de Chim. et de Phys.,' ser. iii., xliii., p. 149), though undertaken for a different purpose, are well adapted to remove all doubt about the very important power possessed by the nitrogenous matter just now alluded to, viz. of maintaining the vital process in the plant, even where the mass of the plant itself receives no increase.

In these experiments lupines, beans, oats, wheat, and cresses were sown in pure pumice-stone dust, washed and burnt, with which was mixed a certain quantity of ash from stable-manure and from seeds similar to those sown. The plants were grown partly under glass bells, with a constantly-renewed supply of air containing carbonic acid. The air supplied and the water used for the plants, were most carefully freed from ammonia. The results of these experiments were as follows:—In an experiment where the plants were grown under a glass bell, 4·780 grammes of seeds (lupines, beans, and cresses), containing 0·227 gramme of nitrogen, gave 16·6 grammes of dried plants; adding the amount of nitrogen in the soil, 0·224 gramme of that element was recovered. In another experiment, where the plants were grown in free atmospheric air, with the exclusion, however, of dew and rain, 4·995 grammes of seeds (lupines, beans, oats, wheat, and cresses) gave 18·73 grammes of dried plants. The seeds contained 0·2307 gramme of nitrogen; the plants and soil, 0·2499 gramme. In the first series of experiments all elements of food were supplied to the plants, except nitrogen; the chief conditions required to form unazotised matter were given, but those required to form azotised matter were altogether excluded.

The growth of a wheat plant in pure water and atmospheric air is unattended with any increase of weight. The normal seed-corn contains a certain quantity of potash, magnesia, and lime, which are required internally for the organic formative process; but it has no excess of those mineral substances that could serve to effect the chemical process of a new production of protoplasm. Where the mineral substances are excluded, the organs will absorb water, but neither carbonic acid nor ammonia; at all events, these two latter substances, even though they be introduced into the plant by means of the water, exert no influence upon the internal process; they suffer no decomposition, and no vegetable matter is formed from their elements.

In Boussingault's experiments, the action of the mineral substances supplied is unmistakable. The weight of the plants produced was nearly 3½ times greater than that of the seeds sown: but the quantity of nitrogenous matter was the same as in the seeds. Hence we have a clear production of non-nitrogenous substance 2½ times more than the original weight of the seeds. A simple calculation shows that the nitrogen in the seed has, under these circumstances, caused the generation of 56 times its own weight of unazotised matter; or, what comes to the same thing (taking the amount of carbon in the latter at 44 per cent. only), the decomposition of 90 times its own weight of carbonic acid.

The course of vegetation in these plants throws sufficient light upon the processes going on in their organism; in the first days their developement was vigourous, afterwards languid. The first-formed leaves withered after a time, and partly dropped off, fresh leaves being developed in their stead, which went on in the same way; and the vegetation seemed to reach a point where the newly developed parts existed at the expense of the decaying portions. A French bean, weighing 0·755 gramme, planted on the 10th May, had by the 30th July developed 17 leaves, of which the first 11 were then dead and gone. The plant flowered, and on the 22nd August, when nearly all the leaves had dropped off, produced a single small bean, which weighed 4 centigrammes, the $\frac{1}{19}$th part of the weight of the seed-bean. The entire crop weighed 2·24 grammes, very nearly three times as much as the seed-bean. In the case of a rye-plant it was very clearly observed how the unfolding of every fresh leaf was attended with the death of one of the old leaves.

In the second series of experiments, the plants had absorbed (from the air) 1·92 milligramme of nitrogen, and produced 0.830 gramme more vegetable substance, giving 43 milligrammes of unazotised matter for every milligramme of nitrogen.

The difference in the developement of a plant in pure water from that of one grown, as in Boussingault's experiments, in a soil supplying the incombustible constituents of food, is clear and unequivocal. The organs first formed received in both cases their elements from the seed; in both, a certain quantity of mineral substances and also of soluble unazotised matter was consumed to form cellulose in the leaves, roots, and stems; and the proportion of the unazotised to the nitrogenous matter was altered. In the plant growing in water, there was a constant decrease of unazotised matter; while in the other a certain quantity of that substance was generated anew. Nothing can be more certain than that in Boussingault's experiments, the first-formed leaves acquired by the supply of mineral substances the faculty of absorbing and decomposing carbonic acid, a power not possessed by the plant developed in pure water ; and that as much soluble unazotised substance was reproduced as had been consumed in the formation of the leaves and roots by the conversion into cellulose of the store originally present.

In the movable constituents of the plant, the relative proportion between the unazotised and the azotised seed constituents was manifestly restored pretty nearly as it existed in the seed; both matters passed through the stem into every new-formed leaf-bud, and took part in the developement of new leaves, by whose operation the consumption of unazotised matter was always made good again within a certain limit, so that the same process could be repeated again and again for months. In every one of the dead leaves (and root fibres) a certain quantity of the azotised substance remained behind, and in the last period of vegetation the floating remainder of this substance was collected in the pod and in the seeds.

The supply of mineral substances had served to effect the continuance of the chemical process, and caused the production of unazotised substances. By the presence of these mineral bodies, and by the cooperation of the azotised matters, new material was engendered from carbonic acid to form the cell-walls, and the term of life was prolonged to its proper limit. The most remarkable point is, that a quantity comparatively so small, of azotised substance derived from the seed, should so long be able to perform its assigned functions, apparently without suffering any alteration ; so that in the body of the living plant, made to produce and collect it, it would seem to possess a kind of indestructibility.

If we consider, that, in the cited experiment with the French bean, a great part of the additional unazotised substances which were produced fell away in the dying leaves from the body of the plant, it will be seen that the supply of mineral substances was of no use to the bean-plant in the absence of nitrogenous food.

Lastly, it is quite intelligible that the amount of azotised matter contained in a bean might perhaps suffice to sustain for years the vegetation of one of the conifers with persistent leaves, and to produce many hundred — perhaps many thousand — times its own weight of woody substance; and that such a plant upon a barren soil altogether unsuited for other plants, might thrive with a very sparing supply of nitrogenous food, if the soil contained a proper store of those mineral substances which are indispensable for the generation of unazotised matter.

The growth of a plant essentially consists in the enlargement and multiplication of the organs of nutrition, i. e. the leaves and roots. The enlargement of the first, or the production of a second leaf or root fibre, requires the same conditions as the production of the first. The analysis of the seeds teaches us with tolerable certainty what these conditions are. In the normal conditions of nutrition, the first roots and leaves, whose elements were supplied by the seed, produce from certain mineral substances organic compounds, which become parts and constituents of themselves, or constituents of fresh leaves and roots, consisting of the same elements and having the identical properties of the first, i. e. they possess the same power to transform inorganic nutritive substances into organic formative materials.

It is quite clear that the enlargement of the first leaves and roots and the production of new ones, must have required azotised and unazotised substances in the same proportion as in the seed, which makes it probable that the organic operations of the plant under the dominion of sunlight uniformly produce in all periods of growth the same materials, i. e. the constituent elements of the seed, which serve to build up the plant, being formed into leaves, stems, and root-fibres, or finally into seed. The soluble constituents of a bud, a tuber, or the root of a perennial plant, are identical with the seed constituents. The cereal plant produces azotised and unazotised substances in the same proportion as in the albumen (farinaceous body). The potato plant produces the constituents of the tuber, which are formed into leaves and branches or roots; or, if the external conditions are no longer favourable to the formation of leaves and roots, accumulate again in the underground stem, to form new tubers.*

While the growth of the plant continues, the first as well as the last leaves and roots will, with a proper supply of food, maintain their existence, since they reproduce out of the nutriment supplied to them the identical constituents from which they themselves arose. The excess of these, which they do not require for their own enlargement, goes to those parts of the plant where the motion of the fluids or the cell-formation is most active,—viz., to the roots, the leaf-buds, or the extreme points of the roots and shoots; and, finally, as in the case of summer plants, to the organs of seed-formation which at the ripening of the seed absorb most of the movable seed-constituents existing in the plant.

The supply of the incombustible elements of food led to the formation of unazotised matter, a portion of which was used to form woody tissue, whilst another portion remained available for the same purpose. The supply of nitrogenous food caused a corresponding production of nitrogenous matter, so that the protoplasm was constantly renewed, and, so long as the chemical process lasted, was increased.

* Boussingault has observed that even seeds weighing two or three milligrammes, sown in an absolutely sterile soil, will produce plants in which all the organs are developed, but their weight, after months, does not amount to much more than that of the original seed, even if they vegetate in the open air; and the result is more marked if they grow in a confined atmosphere. The plants remain delicate, and appear reduced in all dimensions; they may, however, grow, flower, and even bear seed which only requires a fertile soil to produce again a plant of the natural size. ('Compt. rend.' t. xliv. p. 940.)

To enable a plant to flower and bear seed, it would appear necessary in the case of many plants that the activity of the leaves and roots should reach a period of rest. It is only after this that the process of cell-formation seems to gain the ascendancy in a new direction; and the constructive materials being no longer required for the formation of new leaves and roots, are used to form the flower and the seed. In many plants the want of rain, and the consequent deficiency of incombustible nutritive substances, will restrain the formation of leaves and hasten the flowering. Dry, cool weather favours the production of seed. In warm and moist climates the cereals sown in summer bear little or no seed; and on a soil poor in ammonia the root-plants more readily flower and bear seed than on a soil rich in that substance.

If the normal processes of vegetation require a definite proportion of unazotised and azotised materials in the protoplasm which is formed in the plant, it is evident that the want or excess of the mineral substances indispensable for the production of those matters must exercise a very decided influence upon the growth of the plant, and upon the formation of the leaves, roots, and seed. Want of azotised and excess of fixed nutritive substances would lead to the formation of unazotised materials in preponderating quantity; but when these have assumed the form of leaves and roots, they retain a certain amount of nitrogenous matter, thereby impairing the seed formation, a principal condition of which is an excess of protoplasm. An excess of azotised food, with a deficiency of fixed nutritive substances, will be of no use to the plant itself, because the latter can for its organic operations make use of nitrogenous substances only in proportion as they exist in the protoplasm, and the contents of the cell are of no value to the plant in the absence of the materials required to form the cell-walls.

In the process of animal life the organs of the body are constructed from the elements of the egg; the constituent parts of such constructed organs are azotised, whereas in the plant they contain no nitrogen. All processes of vegetative life tend simply to produce the elements of the seed. The plant only lives in generating the egg-constituents and the egg itself; the animal only lives by destroying these very egg-constituents.

On one and the same soil equally suited for the turnip and the wheat-plant, the former produces for the same amount of azotised substance twice as much unazotised matter as the latter. It is manifest that if two plants produce within the same time different quantities of hydrates of carbon (wood, sugar, and amylum), the organs of decomposition must be arranged in a manner not only to afford adequate room for the carbonic acid supplying the carbon, and for the water supplying the hydrogen, as well as to present a suitable extent of surface to the action of the light, but also to permit the liberated oxygen to escape as promptly as it becomes free. If we compare in this respect the leaves of a wheat-plant with those of a turnip-plant, we find a striking difference in their size, and in the amount of water respectively contained in them; and a microscopic examination reveals still greater differences. The wheat-plant has erect leaves, which present to the light a much smaller surface than the leaves of the turnip-plant, which overshadow the ground, preventing the drying of the soil and the exhalation from it of carbonic acid. In the wheat-leaf the stomates are equally thick on both sides; in the turnip-leaf they are much more numerous, although smaller than in the wheat-leaf, and a far greater number of them are found on the lower than on the upper side.

All the facts known respecting the nutrition of plants tend to prove that it is not by a mere osmotic process that they absorb their food, but that the roots perform a very definite active part in selecting from the amount of food presented to them such matters and in such quantities as are best suited to the plant.

The influence of the roots is most manifest in the vegetation of marine and fresh-water plants, whose roots are not in contact with the soil.

These plants received their incombustible nutritive substances from a solution in which these elements are most uniformly mixed and diffused; and yet a comparative analysis of the water and the ash-constituents of these plants shows that each species absorbs from the same solution different quantities of potash, lime, silicic acid, and phosphoric acid.

The ash of duckweed was found to contain 22 parts of potash to 10 parts of chloride of sodium, whereas the water in which the plant had grown contained only 4 parts of potash to 10 parts of chloride of sodium. In the plant the relative proportion of the sulphuric acid to the phosphoric acid was 10 to 14; in the water, 10 to 3.

It is quite the same with marine plants. Sea-water contains for 25 or 26 parts of chloride of sodium 1·21 to 1·35 of chloride of potassium; but the plants growing in it contain more potash than soda. The kelp of the Orkney Islands, which consists of the ashes of many species of fucus,* contains for 26 per cent. of chloride of potassium only 19 per cent. of chloride of sodium.

Sea-water contains manganese, but in such exceedingly small quantity that it would certainly have escaped analysis, were it not invariably found among the ash-constituents of many marine plants. The ash of *Padina pavonia* (a species of tang) is found to contain of this mineral even more than 8 per cent. of the weight of the dried plant.†

By the same power of selection the laminaria withdraw from the sea-water in which they grow the iodine compounds present in it in such exceedingly minute quantities. Chloride of potassium and chloride of sodium have the same form of crystallisation, and so many

* See Gödechen's analysis of the ash of different species of fucus. ('Annal. d. Chem. und Pharm.' liv. 351.)

† To give some idea of the extraordinary power which this plant must possess to withdraw the manganese from sea-water, I need simply state that the quantity of this metal in sea-water is so exceedingly small, that I could find distinct traces of it only by subjecting the sesquioxide of iron, obtained from twenty pounds of sea-water, to a most searching analysis. (Forchhammer and Poggendorff, xcv. p. 84.)

other properties in common, that without the aid of chemical means we cannot accurately distinguish the one from the other. But the plant clearly discriminates between the two salts, for it separates the one from the other, and for every one equivalent of potassium which it absorbs leaves behind in the water more than thirty equivalents of sodium. Manganese and iron, iodine and chlorine, are likewise isomorphous bodies; yet the iodine plant separates one equivalent of iodine in sea-water from many thousand equivalents of chlorine.

The known laws of osmosis, and of the diffusion or interchange of water and salts through a dead membrane or a porous mineral body, give no explanation whatever of the action exercised by a living membrane upon salts in solution, or how they pass through it into the plant. The observations of Graham ('Phil. Mag.' ser. IV. August 1850) show that matters capable of exerting a chemical action upon animal membranes, such as carbonate of potash and caustic potash, causing them to swell and gradually decomposing them, facilitate the passage of water to an extraordinary degree.* Graham remarks that the processes of alteration, decomposition, and new formation, which are incessantly taking place in the membranes and cells in all parts of the plant, and which we have no means of defining or measuring, must entirely change the osmotic process: the permeation of mineral substances through the living vegetable membrane must, therefore, be governed by very complex laws.

Land plants act in the same manner with respect to the soil in which they grow, as marine plants to sea-

* The water in the tubes of his osmometer rose to 167 millimeters, when holding 1/10*m.* per cent. of carbonate of potash in solution; with 1 per cent. of that salt, it rose to 863 millimeters (38 inches, English). In another experiment, the water holding 1 per cent. of sulphate of potash in solution, rose to twelve millimeters; upon the addition of 1/10 per cent. of carbonate of potash to the solution, it rose to 254-264 millimeters; the same potash solution by itself rose only to 92 millimeters. The notion of an osmotic equivalent is altogether inadmissible, if the membrane is chemically altered. Graham's latest investigations on the dialysis of crystalline and amorphous bodies are extremely interesting, and promise to throw considerable light upon the processes in the animal organism.

water. One and the same field presents to the plants growing in it, the alkalis, alkaline earths, phosphoric acid, and ammonia, in absolutely the same form and condition; but the ash of no one species of plant ever shows the same relative proportions of component elements as the ash of another species. Even the parasitical plants, which draw their mineral constituents in a certain state of preparation, from other plants on which they live, as the mistletoe (*Viscum album*), do not comport themselves to the latter as a graftling to a tree, but absorb from the sap very different proportions of mineral constituents ('Annal d. Chem. und Pharm.' liv. 363). Now, as the soil is perfectly passive in respect to the supply of these materials, there must be some agency at work in the plant itself, which regulates the absorption according to the requirements of each plant.

The observations made by Hales (see Appendix C.) show that the exhalation from the surface of the leaves and branches exercises a powerful influence upon the motion of the fluids, and upon the absorption of water from the soil. If the plant drew its mineral food from a solution moving about in the soil and passing immediately into the roots, then two plants of different species or kind, placed in the same conditions, would receive the same mineral substances in the same relative proportions; but, as we have seen, two plants belonging each to a different species contain these substances in the most dissimilar proportions.

That a selection takes place in the absorption of food by the roots is a fact beyond dispute.

In the case of aquatic plants, which grow under water, exhalation is altogether excluded as a possible operating cause of the passing of the food into the body of the plant. In these plants the absorbent surface must exercise very unequal powers of attraction upon the different materials, which are presented by the solution in the same form and in a state of equal mobility; or, what comes to the same thing, the resistance offered to their passage through the outermost cellular layers must be very dissimilar. The case cannot be different with the roots of land-plants, to judge from the unequal proportions of the substances severally absorbed by them.

The power of the roots to preclude the passing of certain substances from the soil into the plant is not absolute. Forchhammer (Poggend. 'Annal.' xcv. 90) detected exceedingly minute traces of lead, zinc and copper in the wood of the beech, birch, and fir; and tin, lead, zinc, and cobalt in that of the oak; but the fact that the outer rind or bark, in particular, is found to contain metals of this kind in perceptibly larger quantities than the wood, clearly points to the accidental nature of their presence, and to their taking no essential part in the vital processes of the plant.

How small the quantities of these metals must be which the roots of these trees absorb may be judged from the fact that hitherto chemical analysis has not been able to detect traces of any other metal than manganese and iron, in the water of wells, brooks, or springs; and their appearance in these wood-plants, which during the growth of half a century or more have absorbed and evaporated an immense quantity of water, is the only proof we possess, that this water must actually have contained these metals in some form or other.

The observations of De Saussure, Schlossberger, and Herth, show that the roots of land and water plants absorb from very dilute saline solutions water and salt in proportions entirely different from those in the fluid; in all cases a greater proportion of water, and a less quantity of salt. In plants watered with very dilute solutions of salts of baryta, Daubeny found no baryta, whereas Knop in similar experiments detected this substance. The general result of all these experiments is that, of themselves, the plants have not the power of offering a permanent resistance to the chemical action of salts and other inorganic compounds upon the exceedingly fine membrane of the root.

Most land-plants in their natural state in the soil can bear no salt solutions, as concentrated as in these experiments, without sickening and dying; and even carbonate of potash and ammonia, which we certainly know to be nutritive substances, act upon many plants as poison, even when present in the water which circulates in the ground only in sufficient quantity to impart a blue tint to red litmus paper. On the other hand, it would be very wonderful if the roots of a plant outside the soil, and in conditions not suitable to their nature should, under the influence of evaporation, be impenetrable for salt solutions.*

Those mineral substances which, like iron, are constant constituents of all plants, though present only in very small proportions, must be regarded very differently from those metals which Forchhammer found in woody plants.

We know the part which iron performs in the animal organism, in which it is present in comparatively no larger quantities than in the seeds of cereals; and we are fully convinced that, without a certain amount of iron in the food of animals, the formation of the blood corpuscles, the agents of one of the chief functions of the blood, is impossible. Hence, by the law of dependence, which links together the life of animals and plants, we are compelled to ascribe to the iron in the plant also an active part in its vital functions so material that the absence of that metal would endanger the very existence of the plant.

Hitherto chemistry has attributed a positive part in the vital process of plants to those incombustible substances only which are common to all, and which differ only in the relative proportions in the plants. But

* If the long limb of a syphon-shaped tube, filled with water and closed with thick pieces of pig or ox bladder tied over both openings, is placed in salt-water or oil, and the other limb is exposed to the air, the water evaporates in the pores of the bladder with which the short limb is closed. By the capillary action of the bladder, the water exuding in gaseous form is taken up again on the other side of the bladder, and a vacuum is thus created in the interior of the tube, whence there is an increased pressure upon the surfaces of both bladders, which forces the salt-water or the oil through the bladder into the tube. ('Researches into some of the Causes of the Motion of Fluids, by J. v. Liebig. Brunswick: Fr. Viewig & Son. 1848.'—p. 67.) A plant in similar conditions is just like a tube closed with penetrable porous membranes.

should the conjecture prove true that iron is a constant constituent of chlorophyll and of the leaves of many flowers, it may be assumed that other metals, found invariably present in certain varieties of plants (as manganese in *Pavonia*, *Zostera*, *Trapa natans*, in many ligneous plants, several cereals, and in the tea shrub), take part in the vital functions, and that certain peculiarities depend upon the presence of those metals. The ash of *Viola calaminaria*, a plant which, in the parts about Aix-la-Chapelle, is held so strongly indicative of the presence of zinc, that the places where it grows are selected for opening new mines in search of zinc ore, is found to contain oxide of zinc. (Alex. Braun.)

As chloride of sodium and chloride of potassium cause some plants to thrive, so iodide of potassium manifestly performs a similar part in others; and if one plant may properly be called a chlorine plant, others may with equal propriety be termed iodine plants, or manganese plants.* (Prince Salm-Horstmar.)

The diversity in the amount of iodine in different varieties of fucus (*Goedechens*), or of alumina in various kinds of *Lycopodium* (Count Laubach), remains, indeed, unexplained; but the power of plants to withdraw substances like iodine, even in the smallest quantities, from the sea water in which they grow, and to accumulate and retain them in their organism, can only be explained upon the assumption that these substances have entered into combination with certain constituent parts of the plants, whereby as long as the plant lives they are prevented from returning to the medium from which they were taken.†

* The examination of the following water-plants revealed the presence of considerable quantities of manganese and iron in their ash, though the water in which they grew apparently contained no trace of manganese:—*Victoria regia* (in the leaf-stalk principally manganese, in the leaf iron); *Nymphœa cœrulea, dentata, lutea; Hydrocharis Humboldti; Nelumbium asperifolium.* (Dr. Zöller.)

† With respect to the copper in the grains of wheat and rye, which Meier of Copenhagen has shown to be a constant constituent of both seeds, Forchhammer (Poggendorff's 'Annal.' xc. 92) remarks:—'It is an old and

It might be supposed that plants become saturated with the substances absorbed from the air and from the soil; and that all materials offered by the soil in solution, or made soluble by the cooperation of the roots, are absorbed without distinction. Upon this assumption, only that substance in the plant could of course pass into it from without, which is withdrawn from the solution within for a formative purpose.

The investigations made by Schultz-Fleeth show that *Nymphæa alba* and *Arundo phragmites* absorb from the same soil and water, the former nearly 13 per cent., the latter 4·7 per cent., of ash constituents; and of these silicic acid in the most unequal proportion; the ash of *Nymphæa alba* containing less than ½ per cent. of that substance, while in the ash of *Arundo phragmites* there are above 71 per cent. Upon the supposition just made, an equal amount of silicic acid is offered to the roots of both plants, and they both take up an equal quantity of it in proportion to the volume of the sap respectively. In the reed plant the silicic acid is incessantly withdrawn from the sap, and deposited in a solid state in the leaves, the margins of the leaves, the sheaths, &c. As the sap within contains less silicic acid than the solution without, fresh quantities of it are absorbed from the latter; but not so with the *Nymphæa*, because the silicic acid taken up by that plant is not consumed in it.

If we accept the same reasons for the passage into the plant of carbonic acid and phosphoric acid, then it can possess no actual power of selection, but the permeation of the nutritive substances will depend upon osmotic conditions.

It certainly cannot be denied that the absorption of nutritive substances depends upon growth or increase

approved practice to steep grains of wheat, intended for sowing, in a solution of sulphate of copper. The usual explanation of this practice is, that sulphate of copper destroys the sporules of blight to which the wheat plant is liable, an explanation which it is not my intention to dispute. Still it might also be held, supposing copper to be an essential constituent of wheat, that the practice in question serves to supply the copper necessary for the vigorous growth of the plant.'

in mass; for as it is certain that a plant will not grow if no food is offered to it, so it is equally certain that it will absorb no nutriment if the external conditions are not favourable to growth. Yet the view given above would force us to conclusions which are not founded in nature; such as, for instance, (1) that there is actually around the roots a solution containing all the ash constituents of the plants; and (2) that the roots of all plants have a similar structure, and their sap is of the same nature.

With regard to the roots, the most common observations appear to show that they possess the power of selecting the proper mineral nutriment for the plant from the matters presented to them. All plants do not thrive equally well in the same soil; one kind succeeds best in soft water, another in hard water, or water abounding in lime; another only on marshy ground; many on fields rich in carbon and carbonic acid, such as the turf-plants; others again on soil containing large quantities of alkaline earths. Many mosses and lichens will grow only on stones, the surfaces of which they sensibly change; others, like *Köleria*, possess the faculty of extracting from silicious sandstone potash and the phosphoric acid so sparingly present in it. Roots of grass attack the felspar rocks, accelerating their disintegration. Rapes and turnips, sanfoin and lucerne, as also the oak and beech, receive the chief part of their food from the subsoil poor in humus; while the cereal and tuberous plants thrive best in the arable surface soil, and in soil abounding in humus. The roots of many parasitic plants are absolutely unable to extract from the soil their necessary food; but this is prepared for them by the roots of the plants on which they grow. Others again, as certain fungi, grow only on vegetable and animal remains, whose azotised and unazotised substances they use for their own construction.

These facts, accepted in their truc significance, seem sufficient to remove all doubt respecting the different action of the roots of plants upon the soil. We know that common *Lycopodium* (club-moss) and ferns absorb alumina; yet we also know that this substance, in the form in which it occurs in all fertile soils, is not soluble in pure water, or water containing carbonic acid; and that it cannot be detected in any other plant growing on the same soil by the side of the club-moss. In like manner, Schultz-Fleeth could not discover in the water in which *Arundo phragmites* (one of the plants most abounding in silicic acid) was growing, sufficient silicic acid to yield a ponderable amount in the composition of 1000 parts of the water.

CHAPTER II.

THE SOIL.

The soil contains the food of plants—Soil and subsoil, conversion of the latter into the former—Power of the soil to withdraw the food of plants from solution in pure and in carbonic acid water; similar action of charcoal; process of surface attraction; chemical decomposition often accompanies this attraction of the food of plants in the soil; general resemblance of the soil in its action to animal charcoal—All arable soils possess the power of absorption, but in different degrees—Mode of the distribution of the food of plants in the soil; chemically and physically fixed condition of the food—Only the physically fixed are available to plants, being made soluble by the roots—Power of the soil to nourish plants; on what dependent—Compo rtment of an exhausted soil in fallow—Means for making the chemically fixed elements of food available to plants—Action of air, weather, decaying organic matters and chemical means—Distribution of phosphoric and silicic acids; influence of organic matters—Action of lime—Process of the absorption of food from the soil by the extremities of the roots—Mechanical preparation of the soil; its influence on the growth of plants; chemical means for preparing the soil—Rotation of crops; its influence on the quality of the soil; action of draining—Plants do not receive their food from a solution circulating in the soil; examination of drain; lysemeter, spring and river water: bog water, food of plants contained in it; Brückenauer spring water contains volatile fatty acids; amount of food of plants in natural waters dependent on the nature of the soil through which they flow—Mud and bog earth as manure; explanation of their action—Manner in which plants take up their food from the soil; experiments on the growth of plants in solutions containing their food; similar experiments with soil containing the food in a physically fixed state—Intimate connection of natural laws—Average crop; necessary quantity of assimilable food in the soil for the production of such; importance of the extent of surface of the food in the soil; the root surface—Quantity of food for a given surface of roots necessary for a wheat or rye crop—Analysis of the soil of a field—Difference between fertility and productive power of a field—Mode of estimating relative extent of root surfaces—Conversion of rye into wheat soil; quantity of food necessary for the purpose; the plan impracticable—Immobility in the soil of the food of plants; experience in agriculture—Real and ideal maximum production—Conversion in practice of the chemically fixed food into an available form—Effect of a manure depends upon the property of the soil—Improper relative proportions of the different elements of food in the soil: effect of this upon the different cultivated plants: means for restoring the proper relative proportions.

FROM the soil plants receive the food necessary for their developement; hence an acquaintance with its chemical and physical properties is important in helping us to understand the nutritive processes of plants, and the operations of agriculture. As a matter of course, a

4

soil to be fertile for cultivated plants, must, as a primary condition, contain in sufficient quantity the nutritive substances required by those plants. But chemical analysis, which determines this relation, gives but rarely a correct standard by which to measure the fertility of different soils, because the nutritive substances therein contained, to be really available and effective, must have a certain form and condition, which analysis reveals but imperfectly.

Rough uncultivated ground, and soil formed from the dust and dried mud of the highroads, are speedily overgrown with weeds, and though often still unfit for the cultivation of cereal and kitchen plants, may yet prove not unfruitful for other plants, requiring, like clover, sanfoin, and lucerne, a large amount of food, and which are often seen thriving luxuriantly on the slopes of railway embankments formed of earth that has never been under cultivation. A similar relation is shown by the subsoil of many fields. In many of them the earth from the deeper layers improves the surface soil, and increases its fertility; in others, the subsoil mixed with the surface soil destroys the fertility of the latter.

It is a remarkable fact that rough uncultivated soil, unsuited for cereal and kitchen plants, may by diligent cultivation during several years, and by the influence of the weather, become fertile enough to produce those plants which it formerly refused to bear. The difference between fertile arable land and barren untilled soil is not the result of any dissimilarity in the nutritive substances which they contain; because in cultivation upon a large scale, to convert the untilled rough soil into fertile arable land, the ground, so far from being enriched, is rather impoverished by the cultivation of other plants on it.

The difference between the subsoil and the arable surface soil, or the crude and the cultivated soil, supposing that both contain the same amount of nutritive substances, can only be founded upon this, that the cultivated ground contains the nutritive substances of plants, not only in a more uniform mixture, but also in another form.

Now as from the influence of cultivation and weather above-mentioned, the rough soil acquires the power of furnishing the elements of food which it contains, in just the same quantity and in the same time as cultivated soil, a power which was formerly wanting in it with regard to certain plants, it cannot be denied that an alteration must have taken place in the original form and fashion of these elements.

Suppose we have a soil consisting of disintegrated rocks: in the smallest particles of such a soil, the nutritive substances of plants, as potash for instance in a silicate, are retained in combination by the chemical attraction of silicic acid, alumina, &c. This attraction has to be overcome by one still more powerful, if the potash is to be liberated and made available for passing into plants. If we find that some plants are perfectly developed in a soil of the kind, which remains unfruitful for others, we are led to assume that the former are able to overcome the chemical resistances opposed to their growth, and that the latter are not. Further, if we find the same soil gradually acquiring the power of producing these latter plants also, we can assign no other reason than this, that by the combined action of air, water, and carbonic acid, aided by mechanical operations, the chemical resistances have been overcome, and the alimentary substances have been reduced to a form in which they are available for absorption even by plants endowed with the feeblest powers of vegetation.

A soil can only then be said to be perfectly fertile for a given species of plant, e.g. wheat, when every part of its horizontal section which is in contact with the roots contains the amount of food required by the plant, in a form allowing the roots to absorb such food at the proper time, and in the proper quantity, during every stage of its developement.

In a former section mention has been made of a property possessed by arable soil, viz. that when brought into contact with solutions of the articles of food most essential for plants in pure water or in water containing carbonic acid, it can withdraw these elements of food from such solutions. This power throws light upon the form and condition in which these materials are contained or combined in the soil.

To estimate this property correctly in its bearing upon the life plants, we must call to mind a similar property in charcoal, which, like arable soil, withdraws from many fluids colouring matters, salts and gases.

This power in charcoal depends upon a chemical attraction proceeding from its surface, and the materials withdrawn from the fluid adhere to the charcoal in exactly the same way that the colouring matter adheres to the fibre of coloured stuffs coated over with it.

The property of decolorising coloured fluids, which animal wood and vegetable fibre share in common with charcoal, is perceptible in those kinds of charcoal only which possess a certain degree of porosity.

Powdered pit coal, and the shining, smooth, blistered charcoal from sugar or blood, have hardly any decolorising action; whereas porous blood-charcoal and bone-charcoal with its fine pores exceed all other varieties in this property.

Among the wood-charcoals, those made from poplar or pine, having wide pores, are inferior to the charcoal of the beech and box tree; all these varieties decolorise in proportion to the extent of surface which attracts colouring matter. The attractive force which charcoal exercises upon colouring matter is about on a par with the feeble affinity of water for salts, which are dissolved by it, but without alteration of their chemical properties. When dissolved in water, a salt simply assumes the fluid state, and its particles acquire mobility; but in all other respects it retains its characteristic properties, which, as is well known, are completely destroyed by the action of a stronger affinity than that of water.

In this respect the attraction of charcoal resembles that of water, for both attract the dissolved matter. If the attraction of the charcoal is somewhat greater than that of the water, then the colouring matter is completely withdrawn from the water; if the attraction of both is equal, a division takes place, and the attraction is only partial.

The materials attracted by the charcoal retain all their chemical properties, and continue unaltered, merely losing their solubility in water; yet very slight circumstances, increasing in the least degree the attractive force of the water, are sufficient again to withdraw from the charcoal the materials absorbed by it, and which simply coat its surface. By a slight addition of alkali to the water the colouring matter may be discharged from the charcoal which has been used to decolorise the fluid, and by treatment with alcohol, the quinine or strychnine absorbed by charcoal from a fluid may be again extracted.

The arable soil possesses, in these respects, the same properties as charcoals. Diluted liquid manure, of deep brown colour and strong smell, filtered through arable soil, flows off colourless and inodorous; and not merely does it lose its smell and colour, but the ammonia, potash, and phosphoric acid which it holds in solution, are also more or less completely withdrawn from it by the soil, and this in a far greater degree than by charcoal. The rocks which by disintegration give rise to arable soil, if reduced to a fine powder, are just as little possessed of this power as pounded coal. On the contrary, contact with pure water or water containing carbonic acid, deprives many silicates of potash, soda, and other constituents, a clear proof that the former cannot possibly withdraw the latter from the water. There is no perceptible connection between the composition of a soil and its power of absorbing potash, ammonia, and phosphoric acid. A soil abounding in clay, with a small proportion of lime in it, possesses this absorptive power in the same degree as a lime soil with a small admixture of clay; but the amount of humus substances will alter the absorptive relation.

By a closer observation we perceive that the absorptive power of arable soil differs in proportion to its greater or less porosity; a dense, heavy clay soil and a loose sandy soil possess the absorptive power in the smallest degree.

There can be no doubt that all the component parts of arable soil have a share in these properties, but only when they possess a certain mechanical condition, like wood or animal charcoal; and that this power of absorption depends, as in charcoal, upon a surface attraction, which is termed a physical attraction, because the attracted particles enter into no chemical combination, but retain their chemical properties.*

The arable soil owes its formation to the disintegration of minerals and rocks, brought about by the action of mighty mechanical and chemical agencies. Though the comparison may not be altogether apt, the rock may be said to stand in about the same relation to the arable soil resulting from its disintegration as the wood or the vegetable fibre to the humus resulting from its decay.

The same causes which in the course of a few years convert wood into humus act also upon rocks, with this difference, however, that it requires the combined action of water, oxygen, and carbonic acid, for probably a thousand years, to produce from basalt, trachyte, felspar, or porphyry, the thinnest layer of arable soil (such as is found in the plains of river valleys and low lands) with all the chemical and physical properties suited for the nutrition of plants. Sawdust possesses the properties of humus no more than powdered rocks have the properties of arable soil. No doubt sawdust may pass into humus and powdered stones into arable soil, but the two states are essentially distinct; and no human art can imitate the operations which were necessary, during immense ages, to convert the divers kinds of rocks into arable soil.

Arable soil, resulting from the disintegration of various kinds of rocks, bears the same relation, in

* The term, 'physical attraction,' as used here, does not signify a peculiar attractive force, but merely designates the ordinary chemical affinity, which shows differences of degree in its manifestation.

respect of absorptive power for *inorganic* substances in solution, as the woody fibre altered by the action of heat bears to *organic* substances in solution.

It has been stated, that from a solution of carbonate of potash or ammonia, or from a solution of phosphate of lime in carbonic acid water, the arable soil will withdraw the potash, ammonia, and phosphoric acid, without any chemical interchange with the constituents of the earth taking place. In this respect the action of arable soil is absolutely like that of charcoal. But it goes farther, for it is sufficiently powerful to sever the connection between the potash or ammonia and the mineral acid, for which they have the greatest affinity, the potash being absorbed by the soil just as though it were not combined with an acid.

In this property arable soil is like animal charcoal, which, by means of the phosphates of the alkaline earths contained in it, decomposes many salts that are not affected by charcoals free from such phosphates; and, without doubt, the lime and magnesia compounds invariably present in arable soil have a share in this decomposing power which it possesses.

We must suppose that the attractive force of the earthy particles would not in itself be strong enough to separate, for instance, potash from nitric acid, and that it requires the additional attraction of the lime or magnesia to decompose the nitrate of potash. On the one side the soil attracts the potash, on the other the lime or magnesia in the earth attracts the nitric acid, and thus the combined attraction effects, as in innumerable instances in chemistry, a separation which could not have been brought about by a simple one.

The process of decomposition effected by arable soil differs only in one respect from the ordinary chemical processes, namely, that in the latter, as a general rule, no soluble potash salt is decomposed by an insoluble lime salt, in such a manner that the potash is thereby made insoluble and the lime soluble. There is evidently here some other attractive force at work, which alters the effect of chemical affinity. If a solution of phosphate of lime in water containing carbonic acid is filtered through a funnel filled with earth, the uppermost layer of the earth first takes up the phosphoric acid or the phosphate of lime from the fluid. Once saturated therewith it no longer stops the free passage of the dissolved phosphate of lime which now reaches the layer beneath; the latter then again becomes saturated in the same way, and thus by degrees the phosphate of lime is completely diffused throughout the earth in the funnel, so that every particle retains on its surface an equal proportion of this substance. If the phosphate of lime were of the colour of madder and the soil colourless, the latter would now actually present the appearance of a madder lake. Just in the same way potash is diffused through the soil when a solution of carbonate of potash is filtered through it; the lower layers receive what the upper do not retain.

There is no need of any special disquisition to show that the phosphate of lime contained in a particle of bone-earth is diffused in exactly the same way through arable soil, with this difference, that the solution of phosphate of lime in rain-water containing carbonic acid is effected at the very spot where the particle lies, and spreads thence downward and in all directions.

The potash and the silicic acid rendered soluble by disintegration, or by the action of water and carbonic acid upon silicates, are diffused through the soil in the same way, so is ammonia also, which is conveyed in rain-water, or is generated by the putrefaction of the azotised constituents in the decayed roots from the successive generation of plants grown on a field.

Every soil must therefore contain potash, silicic acid and phosphoric acid in two different forms, namely, in *chemical* and in *physical* combination: in the one form, infinitely diffused over all the surface of the porous particles of the soil; in the other, in the shape of granules of phosphorite, or apatite and felspar, very unequally distributed.

In a soil abounding in silicate and in phosphate of lime, which has for thousands of years been exposed to the dissolving action of water and carbonic acid, the component particles will be found everywhere physically saturated with potash, ammonia, silicic acid, and phosphoric acid; and it may occur, as in the case of the so-called Russian black-earth, that the phosphate of lime dissolved but not absorbed is deposited again in concretions, or in a crystalline form in the subsoil.

In this state of physical combination the alimentary substances are manifestly in the most favourable condition to serve as food for plants; for it is clear that the roots, in all places where they are in contact with the soil, will find the necessary nutritive substances in the same state of diffusion and readiness as if these substances were in solution in water, but at the same time not movable of themselves, and retained in the soil by so slight a force that the most trifling dissolvent cause brought to bear upon them suffices to effect their solution and transition into the plant.

If it is true that the roots of cultivated plants have no inherent power to overcome the force which retains together potash and silicic acid in the silicates, but that those elements of food only which are in physical combination with the soil can be taken up and made available for nutriment, this explains the difference between cultivated and uncultivated ground, or barren subsoil.

Nothing can be more certain than that the mechanical treatment of the soil and the influence of the weather serve to strengthen the causes which bring about the disintegration and decomposition of the minerals, and the uniform distribution of the elements of food contained in them and rendered soluble. The elements chemically combined in the minerals, are released from that combination, and in the arable soil gradually resulting from this decomposition acquire the form in which they are available as food for plants. It is evident that only by degrees the rough ground can attain the properties of arable soil, and that the time required for this change depends upon the quantity of nutritive substances present, and upon the obstacles which oppose their distribution, or their disintegration and decomposition. The perennial plants, and particularly the so-called weeds, consuming in proportion to the time less food, and absorbing longer, will always thrive on a soil of this description long before annual or summer plants, which in their shorter period of vegetation require a far larger amount of nutritive substances for their full development.

The longer a soil is under cultivation, the more it becomes suited for the growth of summer plants, from the recurrence and operation of the causes by which the nutritive substances are converted from a state of chemical into one of physical combination. To be productive, in the fullest sense of the term, a soil must be able to afford food *at all points* in contact with the roots of the plants; and, however small the quantity of this food may be, it must necessarily be distributed through every part of the soil.

The power of the soil to nourish cultivated plants is therefore in exact proportion to the quantity of nutritive substances which it contains in a state of physical saturation. The quantity of the other elements in a state of chemical combination distributed through the ground is also highly important, as serving to restore the state of saturation when the nutritive substances in physical combination have been withdrawn from the soil by a series of crops reaped from it.

Experience proves that the cultivation of deep-rooting plants, which draw their food principally from the subsoil, does not materially impair the fertility of the surface soil for a succeeding crop of cereal plants; but the successive cultivation of the latter will, in a comparatively small number of years, render the soil incapable of yielding a remunerative crop.

With most of our cultivated fields this state of exhaustion is not permanent. If the ground is left fallow for one or more years, especially if it is well ploughed and harrowed during the time, it recovers the power of yielding a remunerative crop of cereal plants.

Chemical analysis leaves altogether unexplained the

causes of this fact, so highly important to agriculture, and which has been fully established by the experience of several thousand years. If the reason be that cereal plants feed on those substances only which are in physical combination in the surface soil, then we can easily understand the remarkable fact of a field recovering its power of production without any supply of manure; for though the nutriment in this form constitutes but a small portion of the soil by weight, yet it imparts nutritive qualities to a large volume of it; and it is quite intelligible that a soil not originally rich in nutritive substances physically combined, when drained of them by the innumerable underground absorptive organs of a plant, must very speedily become unsuited for the cultivation of that plant.

Now as the cultivated soil is composed in the main of ingredients which are identical with the constituents of uncultivated ground, and as the agencies affecting the decomposition of these ingredients, and the transposition of their constituents affording food to plants are in constant operation, it is easy to conceive how, by the influence of such causes, an exhausted soil, which is in fact nothing else than a soil reduced to its crude state previous to cultivation, must regain the properties which it had lost. With the conversion of a fresh portion of the food elements from a state of *chemical* to one of *physical* combination, the field recovers the power of affording food to a fresh vegetation in such quantity that the crops are again remunerative to the agriculturist.

An exhausted field which is again rendered productive by fallowing, may accordingly be defined as land deficient in *physically combined* nutritive substances necessary for a full crop, while containing an excess of such substances in a *chemically combined* state. The *fallowing season*, therefore, means the time in which the nutritive substances pass over from the one state to the other. It is not the amount of nutritive substances that is increased in fallowing, but the number of particles of their constituents capable of affording nutrition.

What is here asserted of all the mineral nutritive substances without distinction applies equally to every soil constituent required by the plant. The exhaustion of a field may often simply depend upon a deficiency of available silicic acid for the coming crop of cereal plants, while the other food elements may be superabundant.

It is evident from the nature of the process, that if the soil is altogether deficient in disintegrable silicates or soluble earthy phosphates, the action of time, the plough, and the weather in fallow will not restore fertility to a field, and that the effect of disintegrating causes will vary with the time they are in operation, and with the composition of the different soils.

It clearly results from the foregoing observations, that one of the principal requirements of the practical farmer is to know the causes as well as the means whereby the useful nutritive substances present in his field, but not in a form available for nutrition, may be rendered diffusible and capable of doing their work.

The presence of moisture, a certain degree of heat, and free access of air, are the proximate conditions of those changes by which the nutritive substances in chemical combination are made available for the roots. A certain quantity of water is indispensable to transpose the soil-constituents when rendered soluble; water, with the co-operation of carbonic acid, decomposes the silicates, and makes the undissolved phosphates soluble and diffusible through the soil.

The organic remains decaying in the ground afford feeble but long-continued sources of carbonic acid; but without moisture no process of decay can take place. Stagnant water, again, which excludes the access of air, prevents the generation of carbonic acid; and the process of putrefaction is attended with the generation of heat, whereby the temperature of the soil is perceptibly increased.

By the aid of putrescent vegetable and animal remains, a field exhausted by culture will regain its fertility in a shorter time, and the use of farm-yard manure in time of fallow will promote the process. The dense shadow cast by a leafy plant tends to retain moisture longer in the ground, and thus increases the action of the disintegrating agencies during the fallow season.

In a porous soil abounding in lime the putrefactive process of organic matter proceeds much more quickly than in a clay soil; the presence of the alkaline earth, under these circumstances, serving to oxidise the carbonaceous matter, and to convert the ammonia present in the soil into nitric acid.

All kinds of lime, when lixiviated, give up nitrates to the water. Nitric acid is not retained by the porous earth, as is ammonia; but it is carried down combined with lime or magnesia by the rain-water into the deeper layers of the soil. While the formation of nitric acid taking place in the ground is useful for plants which, like clover and peas, draw their food (here including nitrogen) from a greater depth, yet for this very reason fallowing has a less beneficial effect, with a view to the culture of cereal plants, upon a lime soil rich in animal remains; for by the conversion of ammonia into nitric acid, and its removal, the ground becomes poorer in one of the most important elements of the food of plants. The case is conceivable that a field of the kind, if not cultivated for a number of years, may ultimately have its productive powers impaired by a deficiency of nitrogenous food in the soil.

The cause of the exhaustion of a field by the culture of any plant is always, and under all circumstances, dependent upon a deficiency of one or more nutritive substances in those portions of the soil which are in contact with the roots. A field in which these portions are deficient in phosphoric acid in the state of physical combination, will be found unsuited for the production of a proper crop, though it should contain abundance of available potash and silicic acid. The same results will follow from a want of potash, even though phosphoric and silicic acids be plentiful; and equally so from a want of silicic acid, lime, magnesia, or iron, even where potash and phosphoric acid are in abundance.

When the exhaustion of a field is not caused by the absolute deficiency of food elements, when even a more than adequate supply of all the needful nutriment is there, but not in the proper form, and where consequently fallowing will again render the crop remunerative, the farmer has means at his disposal to assist the action of the natural agencies, whereby the conversion of the food elements into the state of physical combination is effected, and thus to shorten the fallowing season, or even in many instances to make it altogether superfluous.

We have seen that the diffusion of earthy phosphates through the soil is effected exclusively by water, which, if containing a certain amount of carbonic acid, dissolves these earthy salts.

Now, there are certain salts, such as chloride of sodium, nitrate of soda, and salts of ammonia, which experience has proved to exercise, under certain conditions, a favourable action upon the productiveness of a field.

These salts, even in their most dilute solutions, possess, like carbonic acid, the remarkable power of dissolving phosphate of lime and phosphate of magnesia; and when such solutions are filtered through arable soil, they behave just like the solution of these phosphates in carbonic acid water. The earth extracts from these salt solutions the dissolved earthy phosphates, and combines with the latter.

Upon arable soil mixed with earthy phosphates in excess, these salt solutions act in the same way as upon earthy phosphates in the unmixed state, that is, they dissolve a certain proportion of the phosphates.

Nitrate of soda and chloride of sodium suffer, by the action of arable soil, a similar decomposition to that of the salts of potash. Soda is absorbed by the soil, and in its stead lime or magnesia enters into solution in combination with the acid.

If we compare the action of arable soil upon salts

of potash and salts of soda, we find that the soil has far less attraction for soda than for potash; so that the same volume of earth which will suffice to remove all the potash from a solution will, in a solution of chloride of sodium or nitrate of soda of the same alkaline strength, leave undecomposed three-fourths of the dissolved chloride of sodium and half of the nitrate of soda.

If, therefore, a field exhausted by culture, which contains earthy phosphate scattered here and there, is manured with nitrate of soda or chloride of sodium, and by the action of rain a dilute solution of these salts is formed, a portion of them will remain undecomposed in the ground, and must in the moist soil exert an influence, weak in itself, but sure to tell in the long run.

Like carbonic acid generated by the putrefaction of vegetable and animal substances, and dissolving in water, these salt solutions become charged with earthy phosphates in all places where these occur. Now when these phosphates diffused through the fluid come into contact with particles of the arable soil not already saturated with them, they are thereby withdrawn from the solution, and the nitrate of soda or chloride of sodium remaining in solution again acquires the power of repeatedly exerting the same dissolving and diffusing action upon phosphates which are not already fixed in the soil by physical attraction, until these salts are finally carried down by rain-water to the deeper layers of the soil, or are totally decomposed.

It is well known that chloride of sodium is present in the blood of all animals, and that it plays a part in the processes of absorption and secretion; hence it may be regarded as indispensable for these functions. We find also that nature has endowed fodder-plants, tuberous and root-plants, which serve more particularly as food for cattle, with a greater power of taking up chloride of sodium from the soil than is possessed by other plants; and agricultural experience shows that the presence of a small amount of common salt is favourable to the luxuriant growth of these plants.

Of nitric acid, it is generally assumed that it may, like ammonia, serve to sustain the body of the plant. Thus, chloride of sodium and the nitrates act in two distinct ways: one direct, by serving as food for the plant; one indirect, by rendering the phosphates available for the purposes of nutrition.

The salts of ammonia act upon earthy phosphates in the same way as the salts just mentioned, but with this distinction, that their power of dissolving phosphates is far greater; a solution of sulphate of ammonia will dissolve twice as much bone-earth as a solution of an equal quantity of chloride of sodium.

However, as regards the phosphates in the soil, the action of the salts of ammonia can hardly be more powerful than that of chloride of sodium or nitrate of soda, since the salts of ammonia are decomposed by the soil much more speedily, and often even immediately; so that, as a general rule, no solution of such a salt can be said to be actually moving about in the soil. But as a certain volume of earth, however small, is required to decompose a given quantity of salts of ammonia, the action of those salts upon this small volume of earth must be all the more powerful. While, then, the action of salts of ammonia is barely perceptible in the somewhat deeper layers of the arable surface soil, that which they exercise on the uppermost layers is so much the stronger. Feichtinger observed that solutions of salts of ammonia decompose many silicates, even felspar, and take up potash from the latter. Thus, by their contact with the arable soil, they not only enrich it with ammonia, but they effect, even in its minutest particles, a thorough transposition of the nutritive substances required by plants.

The vegetable and animal remains in a soil seem to exercise a remarkable influence upon the diffusion of silicates. The experiments made on this point show that the absorptive power of an arable soil for silicic acid is in an inverse ratio to the amount of organic remains in it; so that a soil rich in such remains will, when brought into contact with a solution of silicate of potash, leave a certain amount of silicic acid unabsorbed, whereas an equal bulk of soil poor in organic remains will take up the whole of the silicic acid in the solution. The incorporation of decaying vegetable and animal matter will, therefore, in a soil containing disintegrable silicates, first of all accelerate the decomposition of the silicates, by the action of the carbonic acid generated in the process of decay, and then as these substances diminish the absorptive power of the soil for silicic acid, as soon as this acid has passed into solution, it is distributed through the soil more widely than would have been the case had these substances not been present.

On many fields poor in clay, the growth of grass for several years will, in consequence of the organic matters collecting in the soil, which serve to promote the distribution of the silicic acid, act more favourably on a succeeding crop of a cereal plant than a plentiful application of farm-yard manure, whose organic constituents, quite irrespective of the silicate of potash in the straw, are always in operation to effect the same object. On many other fields, especially on those abounding in lime, where there is no actual deficiency of silicic acid, but the quantity present is not properly distributed through the soil, a dressing of pulverised turf-waste often produces an equally favourable effect on a succeeding cereal crop as a plentiful application of farm-yard manure.

Deficiency or excess of soluble silicic acid in the ground is equally injurious to the growth of cereal plants. A soil which would answer very well for horse-tail or common reed (*Arundo phragmites*, plants abounding in silica) is not on that account equally well suited for the superior kinds of meadow grass, or for cereals, although these demand a rich supply of silicic acid. Such a soil may be improved by drainage, which, by giving free access to air, decomposes and destroys the organic substances present in excessive quantity; or it may derive benefit from a dressing of marl, or of burnt lime, slaked, or fallen to powder by moist air.

Hydrated silicic acid loses its solubility in water by simple drying, and it frequently happens that the drainage of a marshy field will cause the siliceous plants (reeds and horsetail) to disappear. The action exerted upon the soil by hydrate of lime, or by lime slaked or fallen to powder in the air, is twofold. On a soil rich in humus constituents the lime combines, in the first place, with the organic compounds present, which have an acid reaction; it neutralises the acid of the soil, thereby causing the speedy disappearance of many weeds, such as bog-moss (*Sphagnum*) and reed-grasses, which flourish in a sour soil of this kind. Simple contact with acids powerfully promotes the oxidation of metals (copper, lead, iron), while contact with an alkali prevents it (iron coated with a dilute solution of carbonate of soda will not rust). Upon organic substances, the action is the very reverse: acids prevent, and alkalis promote, oxidation or decay. Excess of lime causes the aforesaid destruction of the humose constituents.

In the same degree as the acid humus, by the action of lime, disappears from the ground, the absorptive power of the latter for hydrated silicic acid is increased; and the excess of this acid present loses its mobility in the soil.*

The action of lime, as we see, is so complex, that from its favourable influence upon one field, it is scarcely ever possible to form an opinion of its probable action upon another field, the condition of which is unknown. This is possible only when the causes of its favourable action in the first case are clearly understood.

When lime has improved the condition of a field, simply by neutralising the acid state of the soil, and

* In an experiment made specially for the purpose, it was found that a litre (about a quart) of forest soil, containing 30 per cent. of humose constituents, absorbed from a solution of silicate of potash only 15 milligrammes of silicic acid. But the same soil mixed with 10 per cent. of washed chalk (carbonate of lime) absorbed 1140 milligrammes; and when mixed with 10 per cent. of slaked lime instead of chalk, the absorptive power was increased to such a degree, that a litre absorbed 3169 milligrammes of silicic acid.

destroying the injurious excess of vegetable remains, the farmer will in vain expect a favourable result from the application of lime in the following years, unless the same causes should recur which had originally impaired the fertility of the field.

In a soil wherein there are putrescent and decaying substances not a single plant will thrive, except mushrooms; and it seems that every chemical process going on in the neighbourhood of roots disturbs that of their own. Decaying substances in excess, by generating too much carbonic acid, injure even those plants which thrive particularly well in a humose soil containing a moderate quantity of humus.*

Upon deep-rooting plants, such as turnips, clover, sanfoin, peas and beans, organic matters accumulating largely in the subsoil act very injuriously, especially in clay, where they decay much more slowly than in a lime soil. The process of decay is communicated to the sickening roots, in which spores of fungi find a suitable soil for their developement. When turnips are thus affected, they become the prey of certain insects, which deposit their eggs in the roots, causing in their developement a strange alteration and disturbance of the vegetable process; for in the diseased parts spongy tmuours arise, the inner substance of which becomes soft and emits a bad smell, and in this state serves to nourish the larva of the small fly.

All these processes, however obscure in themselves, are put an end to by applying lime to such a field; a proper lime dressing will always attain this object. Fields that are particularly rich in organic remains

* Gasparini sowed a few grains of spelt in a pot with washed earth from Vesuvius; these produced plants which continued to grow in a healthy state. In another pot, filled with the same earth, he introduced a piece of bread; in this, all the roots in the immediate vicinity of the mouldering bread died away, and the other roots seemed to have turned off towards the sides of the pot. It is clear that spelt would not grow in a soil copiously mixed with bread; and if the decaying roots left by a spelt crop have the same effect, it is not difficult to conceive how the decaying remains which a plant leaves in the ground, may injuriously affect its own growth, or that of other plants. (Russell.)

require a much larger supply of lime than others, to effect their restoration to a healthy state.

It is certain, that in all such cases, the beneficial action of the lime is not attributable to an original deficiency of that body in the soil for plants growing on it; for in that case, considering the rapidity with which it is diffused through the soil, the effect would manifest itself very soon, and even in the course of the first year. But it takes several years before the favourable change in the condition of the soil is effected; proving that the lime operates, not simply as food, but by producing an alteration in the soil, which requires time, that is, a succession of operations.

On a drained marshy soil, in which lime has diminished the excess of hydrated silicic acid, a second application will not produce the same result, because the offensive substances, once removed, will not return; while on a heavy, stiff clay or loam, the application may be repeatedly successful. These kinds of soil are thereby made more friable and richer in available potash. The nature of the change produced is most clearly shown in the hydraulic lime obtained by calcining native cement stones (a hard marl). These cement-stones consist of a mixture of lime and clay, the former being in larger proportion than in calcareous clay soil. After burning, if it is stirred up with a large quantity of water, the separated potash imparts to the fluid all the properties of a weak lye. Clay which before calcination with lime refused to dissolve in acids, is, after calcination, soluble in acids to the whole extent of the silicic acid present.

A calcareous clay soil withdraws from a solution of silicate of potash much less potash after calcination than before, but a much larger quantity of silicic acid.*

Besides the chemical agents mentioned here, which

* At Bogenhausen, near Munich, loam was calcined in the air, and brought into contact with a solution of silicate of potash; before calcination, a litre of this earth took up 1148 milligrammes of potash, and 2007 milligrammes of silicic acid; after calcination, no potash, and 3230 milligrammes of silicic acid.

the farmer may employ to effect the proper distribution of the nutritive substances stored up in his field, and to make the earthy phosphates, the potash, and the silicic acid available to the roots of the plants, he further improves his land by the mechanical operations of agriculture, and by removing from the soil all obstacles that hinder the spreading of the roots, as well as those injurious agencies which interfere with their normal activity, or endanger their healthy condition.

The effect produced by breaking up the ground by the plough, spade, hoe, harrow, and roller, depends upon the fact, that the roots of plants go in search of their food; that the nutritive substances have no locomotion of their own, and cannot of themselves leave the place in which they are. The root, as if it had eyes to see, bends and stretches in the direction of the nutriment; so that the number, thickness, and direction of its filaments indicate the precise spots where they have obtained food.*

The young root forces its way, not like a nail driven with a certain force into a plank, but by the addition of successive layers, which increase its mass from within outwards.

The new substance, which lengthens the extremity of the root, is in contact with the soil. The newer the cells forming at the extremities, the thinner are their walls; as they grow older, the cell-walls thicken, and their outer surface, becoming more woody, is coated in many cases with a layer of corky substance, which, being impenetrable by water, affords, to the soluble matter deposited within, some protection against osmotic influences.

* Pieces of bone are often found completely enclosed by a network of turnip-roots. It is difficult to understand how this could have been accomplished otherwise than by an attraction between the spongioles and the substance of the bone. The cells, or their contents, are incessantly attracted by the fresh surface of a substance, for which the contents have a chemical attraction.

It is owing to this attraction that the roots wind round the piece of bone; they form a root-ball rolled, not from without, but from within, by the new cells constantly formed upon contact with a substance for which they possess a chemical attraction. (Russell.)

Absorption of nutriment from the soil is effected by the extremities of the roots, whose fluid contents are separated from the earthy particles around them by an exceedingly thin membrane alone; and the contact of the two is the more intimate, as the root-fibre during its formation exerts upon the earthy particles a pressure sufficiently powerful, under certain circumstances, to push them aside. The evaporation of water from the leaves produces a vacuum within the plant, whereby a draught is created, which powerfully assists the contact of the moist earthy particles with the cell-wall. The cell and the earth are pressed against each other. Between the fluid contents of the cells and the nutritive substances physically combined in the earthy particles, there manifestly exists a strong chemical attraction, which, with the cooperation of carbonic acid and water, causes the transference of the incombustible matters into the system of the plant.

By the powerful chemical attraction of any body, we understand its entering into a chemical combination, in which it loses its original properties and acquires new ones. In the case of potash, lime, and phosphoric acid, such a combination must take place immediately upon their passage into the cell; for, as already stated, the sap of the roots is always slightly acid. In the sap of the root-shoots of the vine, we can always detect bitartrate of potash; in that of others, oxalate or citrate of potash, or tartrate of lime; but we never find these bases combined in such saps with carbonic acid, nor can phosphate of lime or magnesia be detected. If the fresh sap of the potato-tuber is mixed with ammonia, no precipitate of phosphate of magnesia and ammonia is produced; but this precipitate makes its appearance as soon as the fermentation of the sap has destroyed the (azotised) substance with which the phosphate of magnesia is combined.

Careful mixture and distribution of the nutritive substances present in the soil, are the most important means of rendering them effective.

A piece of bone, weighing half an ounce, placed in

a cubic foot of earth, has no perceptible influence upon its fertility; but when uniformly distributed and physically combined with the minutest particles of the same earth, it attains a maximum of efficacy. The influence of the mechanical operations of agriculture upon the fertility of a soil, however imperfectly the earthy particles may be mixed by the process, is remarkable and often borders upon the marvellous. The spade, which breaks, turns, and mixes the soil, makes a field much more fruitful than the plough, which breaks, turns, and displaces the earth, without mixing it. The effect of both is increased by the harrow and the roller, so that, in the very same places where a crop has grown during the preceding year, a fresh crop will find nutriment; in other words, the earth is not yet exhausted.

The action of chemical agents in distributing the food-elements of plants is still more powerful than that of the mechanical. By applying, in proper quantities, nitrate of soda, salts of ammonia, and chloride of sodium, the farmer not only enriches his field with materials capable of taking part in the nutrition of plants, but he also effects a distribution of the ammonia and potash, thereby replacing or aiding the mechanical work of the plough, and the influence of the weather in the time of fallow.

We are in the habit of calling 'manures' all those materials which, when applied to our fields, increase the crops; but the same effect is produced by the plough. It is evident that the mere fact of a favourable influence exerted by chloride of sodium, nitrate of soda, salts of ammonia, lime, and organic matter, affords no conclusive proof that these have acted as nutritive substances. The work performed by the plough may be compared to the mastication of food by those special organs with which nature has endowed animals; and nothing can be more certain than that the mechanical operations of agriculture do not add to the store of nutritive substances in a field, but that they act beneficially by preparing the existing nutriment for the support of a future crop. With equal certainty we know that chloride of sodium, nitrate of soda, salts of ammonia, humus, and lime, beside the action peculiar to their elements, perform also a kind of digestive function comparable to that of the stomach in animals, and in which they may partly replace each other. These substances, therefore, act beneficially upon those kinds of soil only in which there is a defect, not in the quantity, but in the form and condition of the nutritive elements; and they may accordingly in their permanent action be replaced by a mechanical comminution, or exceedingly fine pulverisation of the soil.

The true art of the practical farmer consists in rightly discriminating the means which must be applied to make the nutritive elements in his field effective, and in distinguishing these means from others which serve to keep up the durable fertility of the land. He must take the greatest care that the physical condition of his ground be such as to permit the smallest roots to reach those places where nutriment is found. The ground must not be so cohesive as to prevent the spreading of the roots.

In a stiff, heavy soil, plants with fine, slender roots will never thrive well, even though the supply of nutritive substances be ample; and in these circumstances, the beneficial influence of green manure and fresh stable dung is unmistakeable. The mechanical condition of the soil is, in fact, altered in a remarkable way by the ploughing in of plants and their remains. A stiff soil loses thereby its cohesion, becoming more friable and crumbling, than it would be by the most diligent ploughing. In a sandy soil, on the other hand, a certain cohesion is hereby produced. Every stem and leaf of the green-manure plants ploughed in, opens up, by its decay, a road by which the delicate roots of the cereals may ramify in all directions to seek their food. Here, too, we must always remember, that the effect calculated to be produced is a question of degree. In many fields, the roots left in the soil of a fine crop of green forage plants will suffice to improve a succeeding cereal crop; and a field from which a crop of lupines has been taken, may possibly give as fine a succeeding cereal crop as a field of equal extent in which the lupines have been ploughed in.

All these observations tend to show the great importance of the mechanical conditions which impart fertility to a soil not originally deficient in the means of nourishing plants; and that a comparatively poorer but well-tilled soil, if its physical condition is more favourable for the activity and developement of the roots, may yield a better harvest than richer land. In like manner, it often happens that the cultivation of a bulbous plant renders the ground better suited for a following cereal, and that a winter crop succeeding a green forage plant, turns out all the better, the richer the previous green forage crop has been, or rather the roots left by it.

Clover and turnips act favourably upon a succeeding winter crop, as their long hardy roots move the subsoil, which is inaccessible to the plough, and open it for the roots of wheat. Here the favourable influence upon the physical condition of the soil far outweighs, for the wheat-plant, the injurious effect of the decrease in the quantity of the chemical conditions resulting from the previous turnip and clover crops. Facts of this nature have but too often misled practical agriculturists to surmise that the physical condition is everything, and that a thorough working and pulverisation of the soil will suffice to command a good crop. These views, however, have always been refuted by time; and all we can consider established is this, that for a series of years the restoration of a proper physical condition in the soil is as important for the productiveness of many fields as manuring, and often more so.

The influence of a proper physical condition of the soil upon the produce can hardly be more convincingly proved than by the facts which agriculture has derived from the drainage of land, under which we comprise the removal of the subsoil water to a greater depth, and the quicker withdrawal from the arable soil of the portion circulating in it. A great many fields unsuited, by their constant humidity, for the cultivation of cereal plants and the superior kinds of forage grasses, have been reclaimed by drainage, and made fit to produce food for man and beast. When the farmer, by means of drainage, keeps within bounds the amount of water in his fields, he controls its injurious influence at all seasons; and by the speedier removal of the water, which soaks the earth and destroys its porosity, a path is opened for the air to reach the deeper layers of the ground, and to exercise upon these the same beneficial influence as upon the surface soil.

In winter, the earth at a depth of 3 to 4 feet is warmer than the external atmosphere; hence the air coming up from the drain-pipes may contribute to keep the temperature of the arable surface higher than it would be without this current of air. The air in the drains is generally richer in carbonic acid than is the case with atmospheric air.

The effect which drainage produces upon the fertility of land may in itself be deemed a proof that plants cannot derive their food from the water moving about in the soil. This view is strongly supported by the analysis of well, drain, and spring water. (See Appendix D.)

The drainage-waters contain all the substances which the rain-water, percolating the surface soil, is capable of dissolving: they contain various salts in trifling proportions, and among these mere traces of potash; ammonia and phosphoric acid are generally absent. In analyses specially made for this purpose, Thomas Way found that in four (drainage) waters no appreciable quantity of potash could be detected in 10 pounds of water; three other waters were found to contain from 2 to 5 pounds of potash in 7,000,000 pounds of water. In three waters no appreciable quantities of phosphoric acid could be discovered: four other waters were found to contain 6 to 12 pounds of phosphoric acid, and 0·6 to 1·8 pounds of ammonia in 7,000,000 pounds of water. In a similar series of analyses, Krocker found that in six drainage-waters no appreciable traces of phosphoric acid or ammonia could be detected; while four other drainage-waters were found to contain not above 2 parts, and two others severally 4 and 6 parts of potash, in 1,000,000 parts of water.

The facts now stated are corroborated by a series of direct and most instructive experiments made by Dr. Fraas, to ascertain what substances the rain falling in the six summer months takes up from the surface soil and carries down into the deeper layers.

In *lysimeters*, or underground rain-gauges specially constructed for the purpose, a collection was made of the rain-water, which trickled through a layer of earth, 6 inches deep by 1 square foot in transverse section, from the 6th April to the 7th October. The rain-gauge kept at a neighbouring observatory indicated, up to the 1st October, a fall of rain amounting to 480·7 millimètres (18·75 inches).*

Four lysimeters were filled with the same earth taken from the subsoil of the stiff clay at Bogenhausen; in two of them (III. and IV.) the earth was manured with 2 pounds of cow's dung; the other two were left unmanured. Nos. II. and IV. were sown with barley.

Calculated upon a square mètre (10·75 square feet) of ground, the following were found to be the quantities of water that had passed through. Dr. Zoeller determined the amount of soluble substances contained in the water; the quantities of phosphoric acid and ammonia were too small to be appreciable:—

* The lysimeter consisted of a square box, open at the top, closed at the bottom; at a depth of six inches from the open top a sieve was inserted, from which, up to the rim, the box was filled with earth. The rain falling upon a square foot of surface, and trickling through the six inches of earth, was collected beneath the sieve, in the box. The box was buried in an open field, up to the border, so that the earth in it was level with the surface of the field. Two lysimeters were filled with lime soils from the banks of the Isar; but one of them broke, and the water could not be collected: hence the results obtained from the other lost their importance as a comparative experiment.

	Lysimeter.			
	I. unmanured: and without vegetation.	II. unmanured: sown with barley.	III. manured: without vegetation.	IV. manured: sown with barley.
Quantity of percolated water	Litres. Pints. 218=383·68	Litres. Pints. 213=374·88	Litres. Pints. 304=535	Litres. Pints. 144=253·5
Quantity of potash contained	Grams. Grains. 0·516=8·0	Grams. Grains. 0·434=6·7	Grams. Grains. 1·265=19·5	Grams. Grains. 0·552=8·5
Or, per hectare, of 2½ acres	Kilog. lbs. avr. 5·16=11·35	Kilog. lbs. avr. 4·34=9·5	Kilog. lbs. avr. 12·65=27·8	Kilog. lbs. avr. 5·52=12·1

In lysimeters I. and II. nearly the same quantities of water percolated through the earth; in the two others the difference is great; the two former alone, therefore, admit of comparison as regards the solvent power of the water.

These experiments show that less than one-half of the rain falling on the field under the given conditions, reached a depth of 6 inches; and that, calculating for 1 million parts of water, the unmanured soils I. and II. gave respectively 2·37 and 2·03 pounds, the manured soils III. and IV. 5·46 and 3·82 pounds of potash. The quantities of potash in the manured soils do not exceed the average quantity of potash found in drainage-water (Krocker).

The barley grown in the earth of lysimeter II. produced, per square mètre, 137·3 grammes (2120 grains) of barley-corns, and 147·9 grammes (2272 grains) of straw, containing in their ashes (the corns in 2·47 per cent., the straw in 4·95 per cent. of ash):—

In the corns . .	0·823 grammes	12·6 grains of potash	
" straw . .	1·410 "	21·8 "	"
Total . .	2·233 "	34·4 "	"

The quantity of potash absorbed by the water from the earth in the first lysimeter, which was not sown with barley, amounted altogether to 0.516 gramme (8·0 grains); in the second lysimeter to 0·434 gramme (6·7 grains). The difference is 0·082 gramme (1·3 grains). If we think ourselves warranted in concluding from this, that the diminution in the quantity of potash in the water of the second lysimeter resulted from its absorption by the roots of the barley, we should be necessarily led to infer that the plants received—

By the agency of the percolating water	0·082	grammes	1·3	grs.
Direct from the soil	2.151	"	33·2	"
Total	2·233	"	34·5	

and, accordingly, 96·4 per cent. direct from the soil, and 3·6 per cent. from the water; that is, 27 times more from the former than from the latter.

Let us now assume, from the results obtained with the third lysimeter, which was filled with earth richly manured with cow-dung, that the rain-water falling on a surface of one hectare (2½ acres) of land, dissolves, out of a layer of arable surface soil 6 inches deep, 12·65 kilogrammes (27·8 lbs.) of potash; and let us compare with this the quantity of potash withdrawn from a hectare of ground by a potato or turnip crop. We know that an average potato crop from a hectare contains in the tubers 204 kilogrammes (449 lbs.) of ash, of which 100 kilogrammes (220 lbs.) are potash; and an average turnip crop, 572 kilogrammes (1258 lbs.) of ash, of which 248 kilogrammes (545 lbs.) are potash; and we easily perceive that, even had the entire amount of the potash dissolved by the rain been conveyed into the plants to serve as food, yet this would be sufficient to supply, with the necessary potash, only the eighth part of the potato tubers and the twentieth part of the turnips severally produced on a hectare of land. The amount of potash in the percolated water shows the quantity of potash which the water could possibly absorb; and as comparatively but a small portion of the percolating water comes in contact with the roots of the plants, and can give up potash to them, it is clear that the constituents of the solution *moving about* in the soil have but a very trifling share in the process of nutrition, while the absence from it of ammonia and phosphoric acid is of itself sufficient to prove that these materials in the soil cannot change their place. The ground must contain a certain amount of moisture to be able to furnish food to plants; but it is not necessary for their growth that the water should be free to move about. It is well known that stagnant water in the soil is injurious to most of the cultivated plants; and the favourable effect upon their growth produced by draining just depends on this, that an outlet is opened to the water moving by the force of its own gravity, and the earth is moistened by that water only which is retained by capillary attraction.

If we regard the porous earth as a system of capillary tubes, the condition which must render them best suited for the growth of plants is unquestionably this, that the narrow capillary spaces should be filled with water, the wide spaces with air, and that all of them should be accessible to the atmosphere. In a moist soil of the kind, affording free access to atmospheric air, the absorbent root-fibres are in most intimate contact with the earthy particles; the outer surface of the root-fibres may here be supposed to form the one, the porous earthy particles the other wall of a capillary vessel, the connection between them being effected by an exceedingly thin layer of water. This condition is equally favourable for the absorption of fixed and of gaseous elements of food. If, on a dry day, a wheat or barley-plant is cautiously pulled up from a loose soil, a cylinder of earthy particles is seen to adhere like a sheath round every root-fibre. It is from these earthy particles that the plant derives the phosphoric acid, potash, silicic acid, &c., as well as the ammonia. These substances are introduced into the plant by means of the thin layer of water, the molecules of which are in motion only in so far as the roots exercise an attractive power upon them.

From the composition of spring-water, and the water of brooks and rivers, every single drop of which has been in contact with rocks, or with the soil of forests and fields, we see what exceedingly minute quantities of phosphoric acid, ammonia, and potash are taken up by water from the earth. In the analysis of water taken from six different springs, Graham, Miller, and Hofmann found no appreciable traces of ammonia and phosphoric acid. In the water of Whitley, there was, in 37,000 gallons (370,000 pounds English), 1 pound of potash, or 1 kilogramme in 135 cubic mètres: just the same in 38,000 gallons from the Critchmere spring; in 32,000 gallons from Velwool; in 145,000 gallons from Hindhead; in 55,000 gallons from the Hasford Mill-brook; and in 17,700 gallons from the spring near Cosford House. The water of the Brunthal spring, near Munich, which is used for drinking in a large portion of the city, contains no ammonia, no phosphoric acid, and in 87,000 pounds, 1 pound of potash.

From these and other analyses of spring, well, and drainage water, we are not warranted in concluding that potash, ammonia, and phosphoric acid are deficient in the water of all springs, brooks, and rivers; on the contrary, it is quite certain that the water in many marshes contains both potash and phosphoric acid in notable quantities.*

The presence of potash, phosphoric acid, iron, and sulphuric acid, in the water of stagnant pools, is easily explained.

* Thus a litre (1·76 pints) of water taken from an artificial pond in the Botanic Garden at Munich, left a residue of 0·425 gramme (6·5 grains), which contained, in 100 parts—

Lime	35·000
Magnesia	12·264
Chloride of sodium	10·100
Potash	3·970
Soda	0·471
Sesquioxide of iron with alumina	0·721
Phosphoric acid.	2·619
Sulphuric acid	8·271
Silicic acid	3·240
Incombustible constituents	76·656
Water lost	23·344
	100·000

In a stagnant pool or bog are gradually collected the remains of dead generations of plants, the roots of which have drawn a quantity of mineral matter from a certain depth of the soil. These vegetable remains undergo decomposition at the bottom of the pools, and their inorganic elements, or ash-constituents, are dissolved by the aid of carbonic acid, and perhaps also of organic acids. They remain dissolved in the water, when the surrounding mud and the earth in contact with this solution have been completely saturated with them.

Scherer found in the three wells at Brückenau all the substances contained in the water above-mentioned, of the Botanic Garden pond, besides acetic, formic, butyric and propionic acids. The mountains all around Brückenau are formed of variegated sandstone (*Bunter sandstein*); the vegetation of the whole surrounding country is most luxuriant, resembling the primeval forests; there are numerous oak-lands and beech-lands, with trees nearly a thousand years old. Hence Scherer is led to attribute the composition of the well-water at Brückenau to the solvent action of rain percolating through a humose soil rich in decaying vegetable substances. ('Annal. der Chem. und Pharm.' i. c. 285.)

It is clear that wherever conditions have been at work similar to those under which the bog-water in the Botanic Garden of Munich and the wells of Brückenau have been formed, the water found on the surface of the earth, in pools, springs, or brooks, will contain in the most varying proportions nutritive elements useful to plants, such as phosphoric acid and potash, which are not found in other waters. In like manner, an arable soil rich in vegetable remains, in which, from the processes of decay incessantly going on, products of an acid character are generated, will be able to give up, to the rain-water percolating through it, phosphoric acid and alkalies, which are thus carried down to the deeper layers, and appear in the drainage water. The quantity of these substances dissolved in the water will depend upon the condition of the soil on which the plants grow, the ash-constituents of which are carried away by the rain-water, from their decaying remains. Where the ground is rocky, covered with a thin coating of earth and a thick clothing of foliage, the water which runs off will carry down to the lower layers all the more fixed elements of vegetable food, in proportion as the layer of earth itself retains less of them. The finer earthy particles of such a soil, washed away by heavy rains, are carried down by torrents to the valleys and low lands, and form a soil of all degrees of fertility according to their chemical condition, which determines their power of absorbing dissolved nutritive substances. But these layers of earth formed from the mud borne down by the torrents will always either be saturated, or gradually become saturated with the nutritive substances contained in the water, from which they are deposited. This, perhaps, explains the difference in the fertilising effects of the waters used for irrigating meadows, which must necessarily vary very much according to the source of the water; that which has collected on hills covered with a rich vegetation, or has been derived from overflowing stagnant pools, will doubtless convey manuring matters to the meadow-lands; whilst water flowing from bare mountains cannot, in this particular respect, exert any action upon the increase of the grass crop. If such increase takes place notwithstanding, the cause must be sought elsewhere.

In many places bog-soil, and the mud from ditches, stagnant waters and ponds, are highly esteemed as fertilising agents; and their influence is explained by the fact, that their smallest particles are saturated with manuring matters, or elements of the food of plants. The same remark applies to the fertility of many tracts of cleared wood-land, where the soil for forty or eighty years, or even longer, has received from the layer of foliage and vegetable remains decaying on it, a certain supply of ash constituents, drawn from a great depth, which are retained by the upper layers of the porous soil, and serve to enrich it.

The injury done to wood-lands by raking away the leaves cannot be explained merely upon the assumption that the soil is deprived of its ash-constituents, which are taken away with the foliage; for, in themselves, the fallen leaves and twigs are poor in nutritive substances, especially potash and phosphoric acid; and besides, these elements do not reach the deeper layers of the soil, where they might be again absorbed by the roots. The injury is, perhaps, rather attributable to the fact, that the remains of leaves and plants constitute a lasting source of carbonic acid, which, carried by rain to the deeper layers, must powerfully contribute to disintegrate and decompose the earthy particles. In a dense wood, where the air is more rarely renewed than in the open plain, this supply of carbonic acid is important; moreover, the thick carpet of leaves protects the ground from being dried by the air, and maintains it in a permanent state of moisture, particularly useful to foliaceous trees, which exhale from their leaves larger quantities of water than the coniferous plants.

To understand the operations of agriculture, it is indispensably necessary that the farmer should have the clearest knowledge of the manner in which plants derive their nutriment from the soil.

The opinion that the roots of plants extract their food immediately from those portions of the soil which are in direct contact with their absorbent surfaces, does not imply that potash, lime, or phosphate of lime, in the solid, undissolved state can penetrate the membrane of the cells;* nor does it imply that the nutritive sub-

* If a glass vessel is filled to the brim with water, in which are a few drops of hydrochloric acid, and covered closely with a piece of bladder, so that the water moistens the bladder and no air is left between them, and the outside of the bladder is carefully dried, it may then be seen how a solid body, without the cooperation of a fluid from the outside, can make its way through the bladder to the water in the glass. For if a little chalk or finely-pulverised phosphate of lime is strewed upon the dried outer surface of the bladder, the powder will disappear in the course of a few hours, and the usual reactions will show the presence of lime and phosphate of lime in the fluid.

Of course the passage of the carbonate and phosphate of lime in the

stances held in solution by the water moving about in the soil may not, under certain circumstances, be absorbed by the roots of the plants. But it is based upon the assumed fact, that the roots receive their food from the thin layer of water which, retained by capillary attraction, is in intimate contact with the earth and with the root surface, and not from more remote layers of water; that between the root surface, the layer of water, and the earthy particles, a reciprocal action goes on, which does not take place between the water and the earthy particles alone. It also assumes as probable, that the nutritive substances adhering, in a state of exceedingly minute division, to the outer surface of the earthy particles, are in direct contact with the fluid of the porous absorbent cell-walls, by means of a very thin layer of water; and that the solution of the solid elements is effected in the pores of the cell-walls, whence they pass immediately into the system of the plant.

The facts in support of this view, briefly recapitulated, are as follow: The roots of all land-plants, and of most marsh-plants, are in direct contact with the earthy particles. These particles of earth have the power of attracting the most important elements of food conveyed to them in watery solution (such as potash, phosphoric acid, silicic acid, ammonia), and of retaining them, just as charcoal retains colouring matters. In most cases that have been investigated it has been found that the water moving about in the ground extracts from the soil scarcely any appreciable quantities of ammonia, no phosphoric acid, and potash in such trifling quantities, that all these together are quite insufficient to afford the requisite supply of these elements to the plants growing in the field.

solid state through the bladder into the water, is only apparent. Both salts are dissolved in the pores of the membrane where they come in contact with the acidulated water, and as the evaporation of the water from the bladder somewhat diminishes the inner pressure as compared to the outer, the stronger outer pressure, assisted by the solvent power of the water, forces the solution inward.

Water stagnant in the ground, so far from promoting the absorption of food, injures the growth of land-plants.

If plants really did receive the elements of their food from a solution which could change its place in the soil, then all drainage waters, spring, brook, and river waters, must contain the principal nutritive substances of all plants; and it must be quite practicable, by continued lixiviation, to extract from every arable soil, without distinction, all the nutritive substances, either entirely, or at least in amount corresponding to the quantity contained in a crop. But, in reality, this is not practicable. By the action of water, the field loses none of the principal conditions of its fertility, in such a degree as perceptibly to impair the growth of plants cultivated on it.

For thousands of years, all fields have been exposed to the lixiviating action of rain-water, without losing their powers of fertility. In all parts of the earth, where man for the first time draws furrows with the plough, he finds the arable crust, or top layer of the field, richer and more fertile than the subsoil. The fertility of the ground is not diminished by plants growing thereon; not until the plants are removed from the ground does it gradually lose its fruitfulness.

The opinion that some cause is at work within the plant itself, which seems to render soluble certain elements of food, and make them available for nutrition, is not contradicted by the experiments of Knop, Sachs, and Stohmann, who have shown that many land-plants, without touching a particle of earth, may be brought to flowering and seed-bearing in water, to which the mineral elements of food have been added. These experiments, which have thrown considerable light upon the physiological importance of the several nutritive substances (see Appendix E.), merely prove how admirably the ground is adapted to the requirements of plants, and how much human ingenuity, knowledge, and minute care, it takes to supply, under circumstances differing so widely from the natural condition, certain properties of arable soil, which insure the healthy growth of plants.

If the supply of nutritive substances in a state of solution were really suited to the nature of the plant and the functions of the roots, it would follow that in such a solution, most abundantly provided with all the elements of food in the most movable form, the plants must thrive the more luxuriantly the fewer the obstacles are which oppose their absorption of food.

A young rye-plant, placed in a fertile soil, will often send forth a bunch of thirty or forty stalks, each of them bearing an ear, and will yield a thousandfold crop of grains, or even more; yet this plant draws its mineral food from a volume of earth, from which the most persevering lixiviation with pure water, or water containing carbonic acid, will not extract even the one-hundredth part of the phosphoric acid and nitrogen, nor the fiftieth part of the potash and the silicic acid, which the plant has drawn from the soil. How is it then possible, under such circumstances, to assume that water alone would have sufficed, by virtue of its solvent power, to render available to the plant all the substances found in it?

None of the plants grown in watery solutions of the mineral elements of their food, even though thriving luxuriantly, will bear the remotest comparison, in the bulk of vegetable matter produced, with plants grown in a fertile soil; and the entire process of developement in them proves that the conditions of thriving growth in the soil are quite of another kind.

The greatest weight of crop obtained by Stohmann from an Indian corn plant grown in water amounted to 84 grammes; while he obtained from another Indian corn plant grown in the soil, at the same time and from the same seed, a crop weighing 346 grammes. In Knop's experiments, the dry weight of two Indian corn plants, the one grown in water, the other in the soil, was found to be as 1 : 7.

The water circulating in the soil contains chloride of sodium, lime, and magnesia—the two latter in combination partly with carbonic acid, partly with mineral acids; and there can hardly be a doubt but that the plant absorbs a portion of these substances from the solution. The same must apply equally to potash, ammonia, and the dissolved phosphates; but the water circulating in the soil, in a normal condition, either does not hold the three last-named substances in solution, or not in sufficient quantities to supply the demands of the plant.

According to the ordinary rules of natural science, when we seek to explain a phenomenon, we leave out of view those cases in which the conditions superinducing the phenomenon are clear and patent. For instance, if we find in bog-water all the ash-constituents of duckweed, there can be no doubt about the form in which they passed into the plant; they were dissolved in water, and they were absorbed in a soluble state. In such a case, we have merely to explain the reason why the several ash-constituents, being all present in one and the same form, have yet passed into the plant in unequal proportions.

If, in another case, we find that the rain-water which falls on a given area of land, dissolves out of the soil many times more potash than was contained in a crop of turnips grown on that area, there is every reason to assume that the turnip, like the duckweed, has absorbed the needful potash from a solution. But, if in the entire quantity of water which falls on the field during the period of vegetation, we find only just so much potash as the turnip crop requires, and no more, the assumption that the potash in the turnips has been derived from this solution would necessarily involve the impossible supposition, that all the watery particles containing potash must have been in contact with the roots of the turnips; otherwise, the latter could not have absorbed so much potash as is actually found in them. This supposition is impossible; because, during the time when the turnip vegetates, there is generally no water circulating in the soil—such, for instance, as might be carried off by drain-pipes.

If the examination of the water in the soil shows it to contain half the quantity of potash required by a turnip crop, there is no need to explain how the dissolved half of the potash has passed into the turnip-plant, but in what form and manner the plant has absorbed the other half deficient in the water.

If, again, by the examination of the water in other fields, we find that it contains only $\frac{1}{4}$; nay, only $\frac{1}{8}$, $\frac{1}{20}$, or $\frac{1}{50}$ of the quantity of potash found in a turnip crop grown upon it; and if we further ascertain that in a soil, favourable for the growth of turnips, the plant always takes up the same quantity of potash from the ground, no matter how much or how little of that substance the water circulating in the soil dissolves from the earth; it follows, that as the water, the soil, and the plant, can alone come into consideration here, the direct power of the water to dissolve potash is of no importance to the plant; and that the plant itself, by the help of water, must have rendered the needful potash soluble.

What is here asserted of one constituent, holds good for all. If, therefore, we find, that by treating a soil with rain-water we can dissolve from it potash, phosphoric acid, and ammonia or nitric acid, in sufficient quantity to account for the presence of these substances in the cereal plants grown on such a soil; while, on the other hand, we find that the plant contains a hundred times more silicic acid than the water could possibly have supplied; the cause of the absorption of silicic acid, which clearly is not in the water, must again here be sought for in the plant itself. Again, if other cases show that an equally abundant crop of corn is obtained on fields, from which water fails to extract phosphoric acid or ammonia, here, too, we are led to the conclusion that the nutritive substances dissolved in the water are of no special importance to the plants in question; but that, as an indispensable requisite, these elements must possess the form most suitable for the action of the root, be this what it may.

The beautiful experiments on vegetation made conjointly by Professor Nägeli and Dr. Zoeller, in the Botanic Garden at Munich, most strikingly prove the correctness of the conclusions to which the analysis of drainage and other waters has led. Instead of growing plants in solutions of the mineral elements of their food, as had been done in all previous experiments, they pursued the very opposite course; they placed the seeds of the plants in a soil containing all the elements of their food in an insoluble state.

In such experiments, it is not easy to find a material which can be used as a substitute for arable soil, and possessing all its properties; and the difficulty is proved by the fact, that none of the plants grown by Boussingault and others, in an artificial soil, abundantly provided with all the elements of food, could even remotely bear comparison with a plant grown in a fertile arable soil. Pulverised charcoal or pumice-stone have the power of extracting many elements of the food of plants from their solutions, and physically fixing them; but they have not, in the moist state, that soft, plastic, and yielding condition of the clay in arable soil, which permits the intimate contact of the roots with the earthy particles. The best substitute for the purpose is coarsely-powdered turf, which, in the moist state, forms a plastic mass, bearing a remote resemblance to clay, and, like arable soil, absorbs all elements of the food of plants from their solutions. Accordingly Nägeli and Zoeller used in their experiments coarsely-powdered turf as the vehicle of the nutritive substances, after having ascertained its absorptive power for the several elements of food.

A litre (1·76 pint) of turf, weighing 324 grammes (4987·6 grs.), was found to absorb from solutions of carbonate of potash, carbonate of ammonia, carbonate of soda, and phosphate of lime—1·45 grammes (22·4 grs.) of potash, 1·227 grammes (19 grs.) of ammonia, 0·205 gramme (3·2 grs.) of soda, and 0·890 gramme (13·7 grs.) of phosphate of lime equal to 0·410 gramme (6·3 grs.) of phosphoric acid.

The quantities of potash and ammonia here given do

not show the total amounts of these substances which the turf will absorb to the point of complete saturation, but merely what it will take up when simply mixed with the solutions, and left in contact with them for a few hours. If we add more of these solutions to the turf-powder, the fluid exhibits an alkaline reaction, which disappears again after one or more days; and it is only at the end of eight days, when the litre (1·76 pint) of turf has taken up 7·892 grammes (121·6 grs.) of potash and 4·169 grammes (64·2 grs.) of ammonia, that the alkaline reaction remains permanent. What we shall hereafter designate as saturated turf contains only $\frac{1}{5}$ of the potash and $\frac{1}{3}$ of the ammonia, which would be absorbed by that substance to the point of complete saturation.

To represent different soils, containing various proportions of nutritive substances, three mixtures were made of saturated and ordinary turf-powder:—

1 mixture contained 1 vol. of saturated turf-powder,
2 " 1 " " and 1 vol. of dry turf-powder,
3 " 1 . " " " 3 " "

These mixtures represented different kinds of earth, in each volume of which the third contained one-fourth, the second one-half the quantity of the nutritive substances present in the first.

The pure turf contained 2·5 per cent. of nitrogen, and 100 grammes yielded 4·4 grammes of ash, which, upon analysis, were found to contain 0·115 gramme of potash, 0·0576 gramme of phosphoric acid, besides lime, sesquioxide of iron, silicic acid, magnesia, sulphuric acid, and soda. (See more fully in Appendix E.)

With each of these mixtures a pot was filled, each pot holding 8½ litres (2592 grammes, = 39917 grs.); a fourth pot, of similar size, contained dry turf-powder.

Taking into consideration the amount of ash in ordinary turf, the four pots severally contained the following quantities of nutritive substances:—

	1st Pot, with common turf.	2d Pot, quarter saturated turf.	3d Pot, half saturated turf.	4th Pot, fully saturated turf.
	Grams. Grains.	Grams. Grains.	Grams. Grains.	Grams. Grains.
Nitrogen...	71· =1093·5	2·60 =40·0	4·32=66·5	8·65=133·2
Potash....	3·18 = 49·0	3·075=47·4	6·15=94·7	12·30=189·5
Phosphoric acid...	1·586= 24·4	0·83 =12·8	1·75=27·0	3·49= 53·8

The figures showing the quantities of nitrogen, potash, and phosphoric acid, express the amount of nitrogen in the dry turf (in the first pot), and the amount of potash and phosphoric acid in its ash. For the other pots, the figures express the quantity of nutritive substances which had been added.

In each of these pots, five dwarf-beans were planted, the weight of which had been carefully determined, and which had been allowed to germinate in pure water.

The plants in the three manured pots grew very evenly, and the luxuriance of their growth excited the astonishment of all who saw them.

During the first month, the plants in pots 2 and 3 (filled respectively with turf ¼ and ½ saturated) presented a finer appearance than the others; but those in pot 4 (filled with saturated turf) soon overtook them; and the difference in the size of the leaves, in proportion to the greater richness of the soil, was very striking.

Remarkable, too, was the influence of the soil upon the term of the vegetating period. Each of the five plants in the pure turf produced a small pod, and, together, the five pods contained 14 seeds. During the ripening of the seeds, the leaves died from below upwards; so that, before the pods had turned yellow, all the leaves had fallen off. The plants in the saturated turf remained green longer than any of the others, and their seeds ripened latest. The last pod of these plants was cropped on July 29, whilst the last pod of the plants in the pure turf had already been cropped on July 16.

The following table shows the crops yielded by all four pots, with the number and weight of the seeds:—

	1st Pot, pure turf.	2d Pot, turf quarter saturated.	3d Pot, turf half saturated.	4th Pot, turf fully saturated.
	Beans.	Beans.	Beans.	Beans.
Number gathered....	14	79	80	103
" sown......	5	5	5	5
	Weight in Grammes.			
	Grammes.	Grammes.	Grammes.	Grammes.
Gathered..........	7·9	56·7	74·3	105·
Sown............	3·965	3·88	4·087	4·055
Excess of crop over seed...........	3·9	52·82	70·213	100·945

What strikes us here at once is the great difference in the number and weight of the seeds respectively gathered from the several pots. The soil richer in nutritive substances yielded not only more, but larger and heavier seeds, the average weight in milligrammes being respectively:—

	1st Pot.	2d Pot.	3d Pot.	4th Pot.
	milligr.	milligr.	milligr.	milligr.
One seed-bean weighed..........	793	776	817	813
One of the gathered beans weighed..	564	718	917	1019

Of the seeds of the plants grown in the first pot (pure turf), seven weighed no more than five of the beans originally sown; whereas those of the plants grown in the saturated turf weighed each 1-5th more than one of the seed-beans.

If we compare the crop of seeds with the quantity of nutritive substances contained in the turf of the four pots, we see at once what influence the form and distribution of the nutritive substances have exercised upon their nutritive power.

The 1-4th saturated turf contained a little above one-half (0·83 gramme) more phosphoric acid than that in the pure turf (1·586 grammes); the potash was doubled; and the amount of nitrogen was increased only by $\frac{1}{27}$th. The crop, however, exceeded that obtained from the plants grown in pure turf, not by $\frac{1}{3}$d (corresponding to the quantity of phosphoric acid added), but it was thirteen times as large. The feeble manuring had caused the turf in the second pot to render thirteen times more nutritive matter for the formation of seed alone, and for the entire plants about thirty times more than the pure turf.

It is evident that only a small proportion of the ash-constituents in the pure turf were present in a form suitable for the nutrition of the bean-plant. They could not be absorbed, because they were in chemical combination in the substance of the turf. To use a somewhat imperfect figure, the nutritive elements in the pure turf may be imagined to be surrounded by the turfy substance, which hinders their contact with the roots; while in the saturated turf these elements form the outer coating of the turfy substance.

The crops of seeds show further that they were not in proportion to the nutritive substances contained in the soil, but that the poorer mixture yielded far more seeds than it should have done in proportion to the production of the richer mixtures. The proportions in the several mixtures were as follows:—

	2d Pot, quar. saturated.	3d Pot, half saturated.	4th Pot, fully saturated.
Amount of manure.......	1	2	4
Crop gathered, as	2	2·8	4

It is not difficult to understand why this should be so. The fact that the $\frac{1}{4}$-saturated turf yielded twice as much crop as corresponded to the amount of manure, proves that the absorbent root-surface had come in contact with double the number of nutritive turf particles.

According to weight, the $\frac{1}{4}$-saturated turf contained, in every cubic centimètre, only $\frac{1}{4}$th of the nutritive substances found in the completely saturated turf; but, by mixing 1 volume of saturated with 3 volumes of unsaturated turf, the former had become far more distributed, and its volume or efficient surface had been made larger. Supposing it were possible to coat 3 volumes of ordinary turf-powder with 1 volume of saturated, so as completely to surround every fragment of the former with saturated turf particles, the bean-plants would, in a soil so prepared, grow as luxuriantly as if every particle of the turf were thoroughly saturated with nutritive substances.

Hence, the higher produce obtained from the comparatively poorer soil proves that it is only the surface of the soil, containing the nutritive elements, which is effective; that the fertility of a soil is not proportionate to the quantity of nutritive substances which chemical analysis proves to be present; and lastly, these facts show that it is not water which, by virtue of its solvent power, has made the nutritive elements available to the roots.

We know by experiment, that when water has dissolved from a saturated soil a certain quantity of ammonia, potash, &c., the same amount of water will not further dissolve from a half-saturated soil (or a soil from which one-half of the absorbed potash and ammonia has already been extracted) half so much as from the saturated soil; but that the earth, in proportion as it has thus become poorer in nutritive substances, will all the more firmly retain the residue of the ingredients absorbed by it.

In the half-saturated turf the nutritive elements are much more firmly bound than in the fully saturated; and, again, in the quarter-saturated much more firmly than in the half-saturated.

Hence, even if the water had been able to dissolve and convey to the roots half as much from the half-saturated as from the fully saturated, and half as much from the quarter-saturated as from the half-saturated, still the produce could not in any case be greater than in proportion to the amount of nutritive substances in the soil. But, in fact, they were far greater, and the roots actually absorbed more nutritive substances than the water could possibly have conveyed to them, even under the most favourable circumstances.

These experiments have, for the first time, afforded direct proof that plants possess the power of absorbing their necessary nutritive elements from a soil in which they are present in physical combination, i.e. in a state wherein they have lost their solubility in water; and the comportment of arable and cultivated soil in general shows that the nutritive substances contained in them must be present in the same form as in the artificial turf soil of these experiments, with this difference, however, that the earthy particles in the arable soil are not merely the vehicles of these substances, but their source. In a soil consisting of turf-powder, a second crop will not succeed so well as the first, unless the nutritive substances which have been removed are again supplied; nor will the soil regain its fertility, however long it be left fallow.

The benefit derived from mechanical tillage of the ground depends upon the law, that the nutritive substances existing in a fruitful soil are not made to change their place by the water circulating in it; that the cultivated plants receive their food principally from the earthy particles with which the roots are in direct contact, out of a solution forming around the roots themselves; and that all nutritive substances lying beyond the immediate reach of the roots, though in themselves quite effective as food, are not directly available for the use of the plants.

There are no isolated laws in nature, but they are all together links in one chain of laws, which are in turn subordinate to a higher and a highest law.

With the natural law, that organic life is developed only in the outermost crust of the earth which is exposed to the sun, is most intimately connected the power of the fragments of that crust which form the arable surface soil, to collect and retain all those nutritive substances on which life depends. A plant is not, like an animal, endowed with special organs to dissolve the food and make it ready for absorption; this preparation of the nutriment is assigned by another law to the fruitful earth itself, which in this respect discharges the functions performed by the stomach and intestines of animals. The arable soil decomposes all salts of potash, of ammonia, and the soluble phosphates; and the potash, ammonia, and phosphoric acid always take the same form in the soil, no matter from what salt they are derived. In performing this function, the plant-bearing earth constitutes for the use of man and beast an immense purifying apparatus, whereby it removes from the water all matters hurtful to the health of animals, and all products resulting from the decay and putrefaction of deceased generations of plants and animals.

The question how much of the several nutritive substances a soil must contain to yield remunerative crops is of great importance, but its exact determination is beset with vast difficulties. If, indeed, the nutritive power of an arable soil depends upon the quantity of substances held in physical combination in the ground, it is evident that a chemical analysis, which cannot rigorously distinguish elements in chemical combination from those in physical combination, must fail to afford any certain conclusion in the matter.

In comparing several equally productive soils, we often find that they differ immensely in their chemical composition; and that of two soils containing, the one 80 to 90 per cent., the other only 20 per cent. of pebbles and sand, the former will frequently yield better crops than the latter. The case is possible, that a soil fruitful in itself may not suffer any diminution of its fertility by being mixed with half its volume of sand, but may actually become more productive, though it now contains, in every part of its transverse section, one-third less nutritive matter than before. The reason is, that by the addition of sand the food-affording surface of the other constituent parts of the soil is enlarged, and on this everything depends as regards the power of the soil to give up to plants the food contained in it.

A soil on which rye thrives well often proves unsuited for the profitable cultivation of wheat, though both plants take from the soil exactly the same constituents.

It is clear that the failure of wheat on such a soil arises from this cause, that the wheat plants, within the allotted period of their existence, do not find nutriment enough for their full developement in the food-supplying soil about their roots, whilst the quantity supplied is ample for the rye plants.

Now chemical analysis proves that such a rye soil altogether contains, to a depth of 5 to 10 inches, fifty—nay, a hundred times more of the food-elements of the wheat plant than would be required for an abundant crop of wheat; and yet, in spite of this superabundance, the field will afford no remunerative crop to the agriculturist.

If we compare the quantities of phosphoric acid and potash drawn from an area of 2½ acres (hectare), by an average wheat crop (2000 kilogrammes=4400 lbs. of grain, and 5000 kilogrammes=11000 lbs. of straw) and a rye crop (1600 kilogrammes=3520 lbs. of grain and 3800 kilogrammes=8360 lbs. of straw), we find that the two crops severally received from the soil—

	Wheat.	Rye.
	Kilogr. lbs. lbs.	Kilogr. lbs. lbs.
Phosphoric acid	25—26= 55 to 57	17— 18= 37 to 39
Potash	52=114	39— 40= 86— 88
Silicic acid	160=352	100—110=220—242

The difference in the absolute requirement is therefore very small. The wheat crop received from the soil only 9 kilogrammes (=20 lbs.) of phosphoric acid, about 12 kilogrammes (=26·4 lbs.) of potash, and 50 to 60 kilogrammes (=110 to 132 lbs.) of silicic acid, more than the rye crop.

Before the true cause was known upon which the nutritive power of arable soil depends, it was utterly incomprehensible how this trifling difference of a few pounds of phosphoric acid, silicic acid, and potash in the requirements of wheat and rye, could make so great a difference in the quality of a field; for in comparison with the total amount of these ingredients actually contained in the rye field, the additional quantity required by the wheat plant is inappreciably small.

This difference would indeed be inconceivable if the nutritive substances required by the cereal plants had any perceptible power of locomotion, for in that case there could not be an actual deficiency of food in any given spot of the soil; every fall of rain would provide the poorer places with nutriment, if the trifling excess required by the wheat above the rye could really be distributed by the agency of water.

Thus, although a soil suited for rye but not for wheat, may contain, within a short distance from the roots of the wheat, a large quantity of phosphoric acid and potash, often amounting, in the volume of earth between two rye plants, to fifty times more than the trifling addition demanded by the wheat, yet, in point of fact, this nutriment cannot reach the roots of the latter.

But if we consider that the nutritive substances cannot of themselves change their place in the ground, the failure of wheat upon a rye field is very simply explained.

If a 2½ acre field yields to an average rye crop (grain and straw) 17 million milligrammes (=37·4 lbs.) of phosphoric acid, 39 million milligrames (=85·8 lbs.) of potash, and 102 million milligrammes (=224·4 grains) of silicic acid, then the rye plants growing on a square decimètre (=15·3 square inches) receive from the soil 17 milligrammes (=0·26 grains) of phosphoric acid, 39 milligrammes (=0·6 grain) of potash, and 102 milligrammes (=1·56 grains) of silicic acid.

Now, from the same area of a good wheat soil, the wheat plants growing on it receive 26 milligrammes of phosphoric acid, 52 milligrammes of potash, and 160 milligrammes of silicic acid. The food-absorbent surface of the rye and wheat plants is not in contact with all the earthy particles which contain food in a square decimètre of the field downwards, but only with a small volume of the soil; and it is quite evident, that if the seed is to thrive in every spot, the earthy particles, which do not happen to come in contact with the roots, must contain as much nutritive matter as the others.

If we could ascertain with any certainty the root-surface which absorbs nutriment, we might infer the volume of earth from which it received food, for every root-fibre is surrounded by a cylinder of earth, the inner wall of which facing the root is as it were gnawed off by the extremities of the root which press downwards, or by the cell-surfaces which are deposited in a downward direction. But in no plant are the diameter and length of the root-fibres determined, and we must rest satisfied with an approximative estimation.

Let us assume that the 17 milligrammes (=0·26 gr.) of phoshporic acid, 39 milligrammes (=0·6 gr.) of potash, and 102 milligrammes (=1·56 grs.) of silicic acid, are absorbed from a mass of earth the transverse section of which is 100 square millimètres (=15·3 square inches), then the rye-field in each square decimètre (10,000 square millimètres) will contain 1700 milligrammes (= 26·2 grs.) of phosphoric acid, 3900 milligrammes (= 60 grs.) of potash, and 10,200 milligrammes (= 15·7 grs.) of silicic acid; that is, a hundred times as much as an average rye crop requires. Now, as the wheat plant, to thrive equally well, must receive half as much again of phosphoric and silicic acid, and 0·4 more potash, from the same portions of the soil, it follows that if a hectare ($2\frac{1}{2}$ acres), to produce an average rye crop, contains

1700	kilogrammes	= 3740	lbs. of	phosphoric acid,
3900	"	= 8580	"	potash, and
10200	"	= 22440	"	silicic acid,

a fertile wheat soil must contain

2560 kilogrammes	= 5632 lbs. of	phosphoric acid,
5200 "	= 11440 "	potash, and
15300 "	= 33660 "	silicic acid,

If a cubic decimètre (1 litre = 1·7 pint) of arable soil weighs on an average 1200 grammes (= 2·64 lbs.), and we assume that the greater number of the roots of a wheat plant do not go deeper than 25 centimètres (10 inches), then the above 1700 milligrammes of phosphoric acid, 3900 milligrammes of potash, and 10,200 milligrammes of silicic acid, must be contained in an available form in 2½ cubic decimètres, or 3000 grammes (= 66 lbs.) of soil: this makes 0·056 per cent. of phosphoric acid, 0·034 per cent. of potash, and 0·34 per cent. of silicic acid.

Before we discuss the inferences which follow from these numbers, we must remember that they involve some hypothetical elements, which ought not to be left out of view. The numbers representing the quantity of ash constituents, which an average rye and wheat crop take from a hectare (2½ acres) in corn and straw, have been determined by chemical analysis, and are not hypothetical. It is therefore certain that a wheat crop draws from the ground half as much again of phosphoric acid and silicic acid, and one-third more potash, than a rye crop.

The supposition that a wheat soil, to the depth of 10 inches, contains in physical combination 0.056 per cent. of phosphoric acid, 0·034 per cent. of potash, and 0·34 per cent. of silicic acid, which makes a hundred times as much as a wheat crop would take in corn and straw from the field, is purely hypothetical; and the present question is to determine the limits up to which this estimate may be accepted as true.

If arable soil is left for twenty-four hours in contact with cold muriatic acid, a certain quantity of potash, phosphoric acid, silicic acid, as well as lime, magnesia, &c. is extracted. If the soil is treated for a long time with *boiling* muriatic acid, the quantities of dissolved silicic acid and potash are much greater. Lastly, by decomposing by fusion the silicates, and then treating with hot muriatic acid, we can obtain all the potash and silicic acid contained in the soil. Without risk of error we may assume that those nutritive substances which can be extracted by cold muriatic acid are most feebly retained by the soil, and approach nearest the elements in physical combination; or, at all events, so near, that by the common disintegrating agencies they very easily pass into this form of combination.

In this way Dr. Zoeller subjected to analysis two kinds of wheat soil—the loam of Bogenhausen and of Weihenstephan, the latter of which in particular represents an excellent wheat soil. One hundred parts of these two soils yielded to cold muriatic acid—

	Phosphoric acid.	Potash.	Silicic acid.
Soil of Weihenstephan	0·219	0·249	0·596
Soil of Bogenhausen........	0·129	0·093	0·674

If these quantities of nutritive elements are present in an available condition in these soils, that of Weihenstephan would contain of phosphoric acid almost 400 times, of potash 700 times, and of silicic acid rather more than 190 times, as much as a wheat crop requires: in the soil of Bogenhausen the amount of phosphoric acid, potash, and silicic acid would be twice as large as the hypothesis presupposed.

The well-known analyses of similar soils by other chemists show that the assumed estimate of the nutritive substances required in a good wheat or rye soil is rather below than above the actual amount; and, in fact, the future prospects of agriculture would be very gloomy, if the ground was not far richer in nutritive substances than has here been hypothetically assumed.

This is, perhaps, the place to state the distinction between the fertility of a field and its productive powers. According to the experiments of Nägeli and Zoeller, mentioned above, the turf may be so saturated with the necessary nutritive substances as to become an extremely fruitful soil for beans; and a comparison of the ash constituents, in the stalks and seeds of the crop, with the quantity which had been added to the turf, shows that the twelve to fourteen-fold quantity of these ash constituents was enough to produce a very abundant seed crop. The porous turf, saturated even in its minutest particles with nutritive elements, favoured in this case an enormous developement of the roots, to which the largeness of the crops is due. Nothing can be more certain than that its power of production measured by time is very small, and that after a very few harvests its fertility would vanish speedily and for ever.

That our corn fields should contain nutritive substances in very great abundance is the necessary condition for a *continuance* of good crops, but it is not indispensable for *one* rich harvest.

A good rye soil is one which produces an average rye crop, but less than an average wheat crop.

From what we have seen, the reason why a wheat plant, which requires from the soil the same elements as the rye plant, will not thrive as well as the latter upon a rye soil, is founded on this, that during the same period of time the wheat needs more of these nutritive substances than the rye, but cannot obtain this additional quantity. Hence, a good wheat soil which yields an average wheat crop, differs from a good rye soil which produces an average rye crop, inasmuch as the wheat soil in all its parts contains more nutritive substances, just in proportion as the wheat crop needs and carries away more than the rye crop.

A good rye soil, which is able to give and does give 1 per cent. of its nutritive substances to an average rye crop, would *necessarily* yield an average wheat crop, if the wheat plants growing upon it could extract 1½ per cent. of its nutriment. But, in fact, this does not take place: whence it follows that the absorbent root-surfaces of the wheat cannot be half as large again as those of the rye; for, were this the case, the roots of the wheat would come into contact with half as many more earthy particles yielding nutriment, i. e. the rye soil would necessarily produce an average wheat crop, which however is not the case.

The comparative returns, in corn and straw, from a rye soil, which has been sown simultaneously half with wheat and half with rye, might therefore enable us to estimate the extent of root surface in wheat and rye plants. If the wheat crop from one-half of such a field, reckoning by the hectare, receives as much phosphoric acid and potash as the rye crop from the other half (17 kilogrammes of phosphoric acid and 39 kilogrammes of potash), this would argue that the roots of the wheat have come in contact with earth yielding as much nutritive substance, and the earth with the same extent of absorbent root surfaces, as in the case of the rye. If the wheat crop contains phosphoric acid, potash, and silicic acid, either more or less than the rye crop, this would lead us to infer a larger or smaller ramification of the roots. Experiments of this kind with rye, wheat, barley, and oats are well worth making, although they have no practical interest for the farmer, but merely a physiological importance, and would finally lead to conclusions, the correctness of which lies within rather wide limits. The absorptive power of the plant, and the time of absorption, make a difference which, however, hereby becomes perceptible.

Of two plants, with the same absorbent root surface, and yielding equal produce, one of which flowers and ripens earlier than the other, the one with the shorter period of vegetation must find somewhat more food, in all the places which furnish its nutriment, in order to receive the same amount as the other, which has a longer time for absorption.

Thus, the only hypothetical assumptions in determining the above numbers are, that the food-absorbent root surfaces of rye and wheat are equal, and that the rye soil yields neither more nor less than exactly 1 per cent. of its nutritive substances. No doubt such a soil has no actual existence; but, supposing that we had such a soil before us, and were to put the question how much nutriment we must add to convert it into a permanently productive wheat soil, the answer would be not hypothetical, but perfectly trustworthy and exact. If

	Phosphoric acid.	Potash.	Silicic acid.
	Kilogr.	Kilogr.	Kilogr.
The wheat soil contains......	2560	5200	15300
The rye soil	1700	3900	10200
The wheat soil is the richer of the two by	860	1300	5100

Hence, to a rye soil of a given condition and productive power, we should have to add, in some form or other, one-half more phosphoric and silicic acid, and one-third more potash, than it already contains, to make it capable of producing average crops of wheat grain and straw.

And to obtain permanently from a wheat soil a crop half as large again as an average harvest, we should add one-half more of nutritive substances than it already contains.

	Phosphoric acid.	Potash.	Silicic acid.
	Kilogr.	Kilogr.	Kilogr.
A hectare of wheat soil contains	2560	5200	10200
One-half more	1280	2600	5100
	3840	7800	15300

These speculations have no other object than to show that a small difference in the absolute quantity of a nutritive element, required by one kind of plant more than by another, presupposes a great excess in the amount of this constituent in the soil. A wheat crop takes from the soil, per hectare (2½ acres), only 8·6 kilogrammes (19 lbs.) more phosphoric acid than a rye crop; but that the wheat-roots may appropriate these 8·6 kilogrammes, the soil must contain a hundred times as much (860 kilogrammes) of phosphoric acid as the rye soil, or perhaps even more.

Although these figures refer to an ideal soil of strictly definite composition, yet the conclusion which we draw is true for all classes of soil.

It is an undoubted fact, that the ground must always, and under all circumstances, contain a larger amount of nutritive substances than the crop grown on it. Supposing the soil to contain, instead of the hundred-fold, only the seventy or fifty-fold quantity of the nutritive elements in the crop, we infer from the law of the immobility of these elements, that, to double the crop, we must add to the field seventy or fifty times the quantity of mineral constituents contained in the produce. In practice the case is different, for no actual field, like our ideal one, contains phosphoric acid, potash, and silicic acid in exactly the relative proportions in which they exist in the ash of rye or wheat. Most fields which are suitable for cereals are fruitful also for potatoes, clover, or turnips, which extract from the soil much more potash than the cereals.

Therefore to convert a rye soil containing more than 3900 kilogrammes of potash, per hectare ($2\frac{1}{2}$ acres), into a wheat soil, would not require an addition of 1300 kilogrammes of potash, but a proportionately less amount would fully answer the purpose.

We shall hereafter discuss at greater length the relations existing between the composition of a soil and its fertility. The main conclusion, which the above figures are intended to illustrate, is the practical impossibility of converting a rye soil into a wheat soil by supplying the deficient ash constituents, or of making a wheat field by the same means produce half as much again as an average crop. Admitting this might be readily accomplished, experimentally, on a small area, yet the price of phosphoric acid, potash, or even of soluble silica, and the impossibility of procuring them for a large number of fields, though in a given field only one of these substances had to be increased in the proportion stated, would oppose insuperable obstacles to the conversion or improvement of land.

The law of the immobility of the mineral elements in the soil explains the agricultural experience of ages, that almost universally, under like climatic conditions, certain fields are suited for certain plants only, and that no plant can be profitably cultivated upon a soil, unless the mineral contents of the soil are in proportion to the special requirements of that plant.

In practice, it is quite impossible, by a supply of mineral substances, to improve the land of an entire country, so that it shall yield crops considerably more abundant than the natural store of food elements in the soil enables it to produce.

Every field has a real and an ideal maximum of productive power corresponding to the nutritive substances which it contains. Under the most favourable cosmical conditions, the real maximum corresponds to that portion of the total amount of nutritive elements, which is present in the soil in an available form, i. e. in a state of physical combination with the soil; the ideal maximum is what might possibly be obtained if the rest of the nutritive substances, which are in chemical combination, were converted into an available form, and distributed through the soil.

Henee, the art of the agriculturist mainly consists in selecting such plants as will thrive best on his land, in adopting a proper system of rotation, and in using all the means at his command to make the nutritive elements in chemical combination available for plants.

The achievements of practical agriculture in these respects are wonderful, and they demonstrate that the triumphs of art far exceed those of science, and that the farmer, by aiding the agencies which improve the chemical and physical condition of his land, can obtain much more abundant crops than by supplying nutritive matters. Because, what he can supply in the shape of manure, with due regard to a proper return, is so small in comparison with the store of nutritive matter contained in a fruitful soil, that a perceptible increase of produce can hardly be expected to result from it.

But what the farmer may achieve by manuring is at best the result—unquestionably a most important one—that his crops suffer no diminution. Where they actually increase, this is less attributable to the addition made to the store of mineral constituents than to their distribution, and to the fact that certain quantities of inoperative substances have been rendered available.

If we wished, by *increasing* the phosphoric acid required for the formation of seed, to enable a wheat-field yielding an average produce of six grains to give two additional grains, it would be necessary to increase by $\frac{1}{3}$rd the whole amount of the phosphoric acid present in the field, and serving for the formation of seed. For it is always but a small fraction of the total quantity supplied that comes into contact with the roots of the plants; and that they may be able to absorb this $\frac{1}{3}$rd more, it is indispensably necessary to increase the phosphoric acid by $\frac{1}{3}$rd in all portions of the soil. This reflection explains the rule found true in experience, that to produce a marked effect upon crops by manuring, a mass of manure must be laid on, utterly disproportionate to the expected increase.

A manure will exercise its beneficial action upon a field in the most marked manner, when it establishes a more suitable relative proportion between the several mineral constituents in the soil; because upon this proportion the crops are dependent. No special argument is needed to demonstrate, that where a wheat soil contains just so much phosphoric acid and potash as will suffice to afford the quantity of these two substances required for a full wheat crop, and no more (accordingly for every part by weight of phosphoric acid two parts by weight of potash), an additional supply of one-half more, or even of double the quantity of potash, cannot exercise the slightest possible influence upon the crop of corn. The wheat-plant requires for its full developement a certain relative proportion of both nutritive substances, and any increase of one beyond this proportion makes the other not a whit more effective, because the additional supply exercises by itself no action.

An increase of phosphoric acid alone has just as little influence in making the returns greater, as an increase of potash alone: this law applies equally to every nutritive substance, potash, magnesia, and silicic acid; no supply of these substances beyond the requirement of the wheat-plant, or its capacity of absorption, will have any effect upon its growth. The relative proportions of the mineral substances, which the plants draw from the soil, are easily determined by analysing the ashes of the produce. It is found by analysis that wheat, potatoes, oats, and clover receive the following proportions of phosphoric acid, potash, lime, magnesia, and silicic acid:—

	Phosphoric acid.	Potash.	Lime and magnesia.	Silicic acid.
Wheat {corn, straw}	1	2·0	0·7	5·7
Potatoes (tubers)	1	3·2	0·48	0·4
Oats {corn, straw}	1	2·1	1·03	5·0
Clover	1	2·6	4·0	1·0
Average	1	2·5	1·5	3·0

Supposing wheat, potatoes, oats, and clover to be cultivated in a field for four years in succession, each of these plants will absorb from the soil the proportion of mineral constituents which it requires; and the sum total divided by the number of years, viz. four, shows the average relative proportion of all the nutritive substances which the soil has lost.

If, in the formula,

Phosphoric acid.		Potass.		Lime and magnesia.		Silicic acid.
n(1·0	:	2·5	:	1·5	:	3·0)

we determine the value of n, which is meant here to designate the number of kilogrammes of phosphoric acid which the four crops have received from the soil, we find for the wheat crop 26 kilogrammes of phosphoric acid, for the potato crop 25 kilogrammes, for the oat crop 27 kilogrammes, and for the clover crop 36 kilogrammes—altogether, 114 kilogrammes; multiplying the above proportional numbers by this number, we obtain the sum total of all the nutritive substances extracted from the soil by the four crops.

With the help of these proportional numbers, we are better able than before to give some more accurate explanations.

Suppose that the soil of a certain field contains, in an available state, the requisite quantities of phosphoric acid, potash, lime, and magnesia, to supply the four crops stated above, but that it is deficient in the proper proportion of silicic acid—containing, for example, for 1 part by weight of phosphoric acid, only 2½ parts of silicic acid, in an available condition—this deficiency will, in the first place, be felt in the crops of cereal plants, whilst the potato and clover crops, on the contrary, will not be at all diminished. It will depend upon the weather to determine whether this deficiency in the crop of cereal plants extends both to corn and straw or is confined to the straw alone. A want of potash, in proportion to all the other constituents, will barely affect the wheat and oat crops, but it will reduce the potato crop; in like manner, a want of lime and magnesia will impair the clover crop.

If the ground can furnish one-tenth more potash, lime, magnesia, and silicic acid than corresponds to the given proportion of phosphoric acid—thus, if,

	Phosphoric acid.	Potash.	Lime and Magnesia.	Silicic acid.
Instead of	1	2·5	1·5	3·0
The ground should be able to furnish...........	1	2·75	1·65	3·3

the crops would not turn out larger than before. But if, in such a field, the quantity of phosphoric acid is increased, the produce will increase, until the right proportion is restored between the phosphoric acid and the other mineral constituents. The additional supply of phosphoric acid serves in this case to increase the amount of potash, lime, and silicic acid in the produce; but if this additional supply exceeds one-tenth of the phosphoric acid present in the soil, the quantity in excess remains ineffective. Up to this limit, every pound —nay, every ounce—of phosphoric acid supplied has, in this case, a fully determinate action.

If potash or lime alone is wanted to restore the right proportion among the nutritive substances in the soil, a supply of ash or lime will increase the produce of all the crops—the additional supply of lime effecting, in this case, an increase in the amount of phosphoric acid and potash in the augmented produce.

If we find that a soil will not bear a remunerative crop of cereal plants, though it remains fruitful for other plants, such as potatoes, clover, or turnips, which require just as much phosphoric acid, potash, and lime, as the cereals, we may assume that the soil had the latter substances in excess, but was deficient in silicic acid. And if, in the course of two or three years, during which other produce is cultivated on it, the land recovers its fertility for cereals, this must be because it contained, though unequally divided and distributed, an excess of silicic acid also, which, during the fallow season, migrated from the places where it was in excess to those where it was deficient; so that when the subsequent period of cultivation began, there was in all these places the right proportion of *all* the nutritive substances needed by cereal plants.

For similar reasons, if peas or beans can be cultivated on a given field only at certain intervals, and experience shows that skilful, industrious tillage is usually more effective than manure in shortening these intervals, we may infer that in such cases the nutritive substances were not deficient in total quantity in the whole field, but in proper proportion in all parts of the field.

CHAPTER III.

ACTION OF SOIL ON FOOD OF PLANTS IN MANURE.

Manures: meaning of the term; their action as food of plants and means for improving the soil—Effect on soils with different powers of absorption—Each soil possesses a definite power of absorption; the distribution of the food of plants in the soil is inversely to the power of absorption; means of counteracting the absorptive power—Absorption number, notion of; comparison of in different fields; its importance in husbandry—Soil saturated with food of plants; its comportment with water—Quantity of food to saturate a soil—A saturated soil not required for the growth of plants—Manuring may be compared to the application of earth saturated with food—Importance of the uniform distribution of food in manures; fresh and rotted stall manure; compost; importance of powdered turf for the preparation of manure—Quantity of food in unmanured fields and their powers of production; increase of the latter apparently out of proportion to the manure added; experiments on this point; explanation; composition of the soil and its absorptive power compared with the requirements of the plants to be cultivated on it; surface and subsoil plants, the tillage and manuring respectively required by each—Clover sickness; experiments of Gilbert and Lawes; their conclusions; value of them.

THE term 'manure' is commonly used to designate all matters which, applied to a field, will increase the amount of its future produce, or, when the land has been exhausted by cultivation, will restore its capability of yielding remunerative harvests.

Manuring agents act partly in a direct manner as elements of food, and partly, like common salt, nitrate of soda, or salts of ammonia, by enhancing the effect of the mechanical operations of tillage, so that they frequently exert as favourable an influence as the actual increase of the nutritive substances in the ground.

Of the two last-named compounds, nitrate of soda contains a nutritive substance in the nitric acid, and salts of ammonia in the ammonia. Hence it is extremely difficult in individual cases to determine whether their action is due to their nutritive constituents, or to the fact that they have brought about the absorption of other nutritive substances.

In a fertile soil tillage and manuring have a definite relation to one another. If, after a rich harvest, the field is prepared by tillage alone to produce a similar rich crop in the next year, that is, if the mechanical means are sufficient to distribute the store of nutritive substances so uniformly that the plants of the following season will find as much nutriment in all parts of the soil as during the last, any further supply of mineral constituents by manuring would be mere waste; but, where a field is not in that condition, the deficiency must be supplied by manure, in order to restore the original power of production. Thus, in a certain sense, the mechanical operations of tillage and of manure are supplementary to one another.

Of two similar fields, manured in exactly the same way, if the one has been well tilled, and the other badly tilled, the former will yield a richer crop, i. e. the manure seems to have a better effect upon this than upon the badly tilled field.

If one of two farmers knows his land better, and cultivates it more judiciously than the other, the former will, in a given time, obtain as good crops with less manure, or richer crops with the same quantity of manure.

All these facts should be considered in estimating the value of manuring agents; but, as science has no standard for measuring the results of the mechanical operations of tillage, this cannot be taken into account here, and we must confine ourselves to that which can be scientifically measured and compared.

When two fields are equally rich in nutritive substances, it often happens that the one, by tillage alone, or by tillage combined with manuring, will be brought much sooner than the other into a condition to yield a succession of remunerative crops of cereal or other plants.

On a light sandy soil, all kinds of manure act more rapidly and effectively than on clay. The sand is more grateful, say the farmers, for the manure bestowed upon it, and yields a more abundant return than other soils for what it has received. The nitrogenous manures, such as wool, horn-shavings, bristles, and blood, which, as we know for a certainty, act by the formation of ammonia, frequently exercise a far more favorable influence upon many plants than ammonia itself. In other cases, bone-earth acts more powerfully upon the future crop than superphosphate of lime; and sometimes ash will prove more fertilising than if the amount of potash contained in it were directly laid upon the field.

All these facts are most intimately connected with the faculty of arable soils to extract or absorb phosphoric acid, ammonia, potash, and silicic acid from their solutions. The restoration of the productive power to an exhausted field by the mechanical operations of tillage and fallowing alone, without manure, presupposes that in certain parts of the field there must have been an excess of nutritive substances which dispersed in the soil and extended to other places where such substances were deficient.

This distribution demands a certain time. The excess of nutritive elements must first be dissolved, that they may be able to move towards those parts which have lost their elements of food by a previous harvest. The closer these superabundant deposits lie to each other, the shorter is the way over which the substances have to travel; and the less the absorptive power of the intervening earth particles for these nutritive substances, the more speedily will the productive power of the soil be restored.

Every arable soil possesses, for potash and the other substances mentioned, a determinate power of absorption, which may be expressed by the number of milligrammes absorbed by 1 cubic decimetre (= 1000 cubic centimetres) of earth. Thus, for instance:—

Cubic decimet.		Cubic inches.					Milligrammes.		Grains.	
1	=	61	of lime soil	from	Cuba	absorbed	1360	=	21	potash.
1		"	loam	"	Bogenhausen	"	2260	=	35	"
1		"	soil	"	Weihenstephan	"	2601	=	40	"
1		"	soil	"	Hungary	"	3377	=	52	"
1		"	garden mould	"	Munich	"	2344	=	36	"

It will be seen at once that these differences in absorptive power are very considerable. One volume of earth from Weihenstephan absorbs nearly twice as much potash as an equal bulk of soil from Cuba; the Hungarian earth, here examined, absorbs 2½ times as much.

These figures show that a certain quantity of potash, say 2600 milligrammes, if supplied to the Weihenstephan soil, will spread in a space of 1 cubic decimetre of earth. If we were to pour the potash, in solution, on a small plot of ground, 1 square decimetre in area, the potash would penetrate to a depth of 1 decimetre (= 3·94 inches), and no deeper; every cubic centimetre (= ·061 cubic inch) would receive 2·6 milligrammes (= ·04 grain) of potash, but the layers beneath would receive none, or at least no appreciable quantity of it.

If the same solution were poured on an equal area of Hungarian or Cuban soil, the potash filtering through would penetrate, in the former, to a depth of somewhat above 7 centimetres (= 2·7 inches); in the latter, to a depth of 19 centimetres (= 7·5 inches).

The diffusibility of potash in a soil is in an inverse ratio to the absorptive power of that soil; half the absorptive power corresponds to double the diffusibility. In a similar way potash will spread in a field during the time of fallow. From the spot where the potash is set free from a silicate by disintegration, it will diffuse itself through a volume of earth so much the larger in proportion as the absorptive power of the earth for potash is smaller.

The absorptive power of arable soil for silicic acid differs just as much as for potash.

Thus from a solution of silicate of potash, 1 cubic decimetre (= 61 cubic inches) of these different soils absorbed the following quantities of silicic acid:—

Forest soil.	Hungarian.	Garden mould I.	Bogenhausen.	Garden mould II.
Milligr. Grains.	Milligr. Grains.	Milligr. Grains.	Milligr. Grains.	Milligr. Grains.
15=0·23	2644=43·8	2425=37·3	2007=31	1085=16·7

Whence to express the relative diffusibility of silicic

acid in these soils, we have the following proportion:—

Hungarian.	Garden mould I.	Bogenhausen.	Garden mould II.	Forest soil.
1·0	1·09	1·31	2·43	17·6

The same quantity of silicic acid which would saturate 1000 cubic centimetres of Hungarian earth, would furnish a maximum supply for 1311 cubic centimetres of Bogenhausen loam, 2430 cubic centimetres of garden mould II., and 17,600 cubic centimetres of forest soil.

Ammonia, in the pure state, or in the form of salts of ammonia, is absorbed by arable soil just in the same way as potash: one kilogramme (= 2·2 lbs.) of the following earths will absorb respectively these quantities of ammonia:—

Cuban.	Schleissheim.	Garden mould.	Bogenhausen.
Milligr. Grains.	Milligr. Grains.	Milligr. Grains.	Milligr. Grains.
5520=85	3900=60	3240=49·9	2600=40

which gives the following numbers for the relative diffusibility of ammonia:—

Cuban.	Schleissheim.	Garden mould.	Bogenhausen.
1·0	1·24	1·50	2.12

The absorptive power of arable soils for phosphate of lime, phosphate of magnesia, and phosphate of magnesia and ammonia, may be determined in the same way, and the relative diffusibility of these several constituents in different soils may be expressed numerically.

By the term 'absorption number,' we designate, in the following pages, the quantity reckoned in milligrammes (= 0·0154 grain) of the several mineral constituents, which one cubic decimetre (= 61 cubic inches) of earth extracts from their solutions.

To determine the condition of a field, the action of the manures applied to it, and the depth to which the several nutritive substances will penetrate, it is important to establish proportionately the absorptive power of the soil for each of them; thus, for example, 1 cubic decimetre of Bogenhausen loam absorbs:—

	Ammonia.	Phosphate of Magnesia and Ammonia.	Potash.	Phosphate of Lime.
	Milligrammes.	Milligrammes.	Milligrammes.	Milligrammes.
	2600	2565	2366	1098
Relative diffusibility	1·0	1·01	1·10	2·36

Accordingly, the second series of these numbers expresses that if a certain quantity of ammonia in its passage through the soil penetrates to a depth of 10 centimetres, the same quantity of potash will attain a depth of 11 centimetres, and a like quantity of phosphate of lime will reach 23·6 centimetres.

In a soil like the Bogenhausen, which absorbs per cubic decimetre 1098 milligrammes of dissolved phosphate of lime, let us suppose that granules of phosphate of lime are dispersed, and that in one spot of the ground one of these granules weighing 22 milligrammes ($\frac{1}{3}$ of a grain) during the course of a certain time becomes soluble in carbonic acid water, and spreads in the surrounding soil; first of all the earth immediately around this granule will be saturated with phosphate of lime, then as the carbonic acid remains in the water and its solvent power continues, a fresh solution will be formed, which will again offer phosphate of lime for absorption to a wider extent of earth; at length, when the 22 milligrammes of phosphate of lime are thoroughly diffused in the surrounding earth, they will supply 20 cubic decimetres of earth with the maximum of this nutritive substance in the form best suited for absorption. The rapidity with which the phosphate of lime will dissolve and spread depends upon its extent of surface; accordingly, if we suppose the granule to be converted into a fine powder, a solution will be formed richer in phosphate of lime just in proportion to the greater number of particles exposed within the same time to the solvent action of the carbonic acid. Therefore, assuming that in a certain state of greater division twice or three times as much is dissolved in a given time, we infer that distribution under favourable circumstances will take place in one-half or one-third of the time it would take without the division.

If, therefore, in a given case the restoration of the productive power in a soil by fallowing or manuring depends upon the earth when drained of phosphoric acid by the roots of plants receiving the needful phosphoric acid back again from the surrounding earthy particles, it follows that with an equal amount of earthy phosphates the time required to accomplish this end will be shortened in proportion to the division of the earthy phosphates.

Straw manure, after decay, leaves silicate of potash behind, and in the process of putrefaction evolves carbonic acid, which by its action upon the silicates sets free silicic acid; hence by using this manure the diffusion of silicic acid must be promoted as the organic matters absorb none of it, and they, when mixed with the earth, must diminish the absorptive powers of the soil.

The forest soil above mentioned absorbed only very small quantities of silicic acid from its alkaline solutions; and it is evident that the addition of such soil to the Hungarian earth would have the effect of diffusing through a larger volume of earth the silicic acid set free by disintegration.

It is not, however, the case with every soil, that its absorptive power for silicic acid decreases in equal proportion to the quantity of combustible substances which it contains. Thus the Hungarian earth above alluded to contains more (9·8 per cent.) combustible matter than the Bogenhausen loam (8·7 per cent.), yet its absorptive power for silicic acid is not less but greater than that of the latter. Hence it follows that there are other circumstances which influence the absorptive power of the soil, and consequently the diffusibility of silicic acid. A soil abounding in hydrated silicic acid will, under any circumstances, absorb less silicic acid than one deficient in that acid, even though the latter soil should contain a much larger amount of organic substances.

The 'absorption numbers' of two different arable soils afford no criterion for determining the quality of the soil or the amount of nutritive substances which it contains; they merely tell us that, in the one soil, the elements of the food of plants will spread beyond certain places, further than in the other; that the one soil opposes greater obstacles to their diffusion than the other. The farmer, in learning the strength of these obstacles, finds out by experience whether they exert a beneficial or adverse influence upon the cultivation of his fields, and ascertains the means of removing the injurious or strengthening the beneficial influences.

On comparing a fruitful sandy soil with an equally fruitful loam or marl, as regards the nutritive substances contained in them, we are surprised to find that the sand with one-half or even one-fourth, of the total substances contained in the loam, will furnish an equally rich harvest. To understand this circumstance properly, we must remember that the nutrition of a plant depends less upon the quantity, than upon the form of the nutriment in the soil; just in the same way as, for example, half an ounce of animal charcoal presents as large an acting surface as a pound of wood charcoal. If the smaller quantity of nutritive substances in the sandy soil presents as large a surface for absorption as the larger quantity of those substances in the loam, the plants must thrive as well upon the former as upon the latter.

If a cubic decimetre of a fruitful loam is mixed with 9 cubic decimetres of silicious sand, so that every particle of sand is surrounded with particles of loam, as many root-fibres and particles of loam will come into contact in the mixed as in an equal volume of the unmixed soil; and if all the particles of loam can yield the same nutriment, plants will receive from the mixed just as much as from the unmixed soil, though, on the whole, the latter is ten times richer.

All fruitful sandy soils consist of a mixture of sand

with more or less clay or loam; and as silicious sand has a very limited power of absorbing potash and the other mineral constituents of plants, the ingredients of the supplied manure, which have become soluble, spread sooner and penetrate deeper into a sandy soil, which also gives back comparatively more of them than any other soil. In many cases, therefore, a stiff loam may be improved by sand; as, on the other hand, the addition of loam to a sandy soil will cause the nutritive substances, supplied by the manure, to remain nearer the surface or to be retained more firmly in the arable top layer.

But as a sandy soil gives up at harvest more nutritive substances in proportion to what it contains, than a fruitful loam, a more speedy exhaustion is the consequence; its power of production does not last long, and can only be sustained by frequent manuring, to supply the constituents which have been removed. Exactly in the same degree, as the manure acts more beneficially in restoring the productive power, the effect of the mechanical operations of tillage becomes less marked.

The same causes which restore to an exhausted loam a large portion of its lost productive power, if the land is but sufficiently broken up by the plough, are at work in a sandy soil also; but they produce little or no result, because the sand is deficient in those substances which the action of the plough is intended to render available.

As the surface of a hectare (2½ acres) represents 1 million square decimetres, the absorption numbers express the number of kilogrammes of potash, phosphoric acid, and silicic acid, which, when applied on a field, will spread from the surface downwards to a depth of 10 centimetres (about 4 inches). Völker, Henneberg, and Stohmann, in experiments made upon different soils to determine their absorption numbers for ammonia, observed that the earth retained a greater quantity from a concentrated than from a dilute solution of ammonia or salts of ammonia; whence it follows, as a matter of course, that the ammonia is divided between the water and the soil, and that from a soil fully saturated with ammonia, pure water will extract a certain quantity of it; just as charcoal will completely withdraw the colouring matter from a slightly coloured fluid, but from one more deeply coloured will extract a much larger quantity; a part of which, however, is but feebly combined and may be removed by water.

In Völker's experiments, treatment with a copious amount of water extracted one-half the ammonia from a soil saturated therewith; the other half was retained by the earth.

Soils which contain much decaying vegetable matter absorb more ammonia and retain it more firmly than soils that are poorer in decaying organic substances. Even assuming that two cubic decimetres of earth, instead of one, are required to retain completely the amount of ammonia indicated by the absorption number, it is clear that ordinary manuring with an agent abounding in ammonia, such as guano or salts of ammonia, can enrich the earth with this substance only to a very inconsiderable depth.

To saturate with ammonia, a hectare (2½ acres) of Bogenhausen loam, from the surface downwards to the depth of one decimetre, fully, or to half-saturate it to the depth of two decimetres (7·8 inches), would require a supply of 2600 kilogrammes or 52 cwts. of pure ammonia, or 200 cwts. of sulphate of ammonia.

If 800 kilogrammes of guano, containing 10 per cent. of ammonia, are applied to a hectare of Bogenhausen soil, the amount of ammonia added is 80 kilogrammes (= 176 lbs.), which is a little more than the thirtieth part of the quantity required to half-saturate the soil to a depth of 20 centimetres. Without the plough and harrow, the quantity of ammonia contained in the guano would not penetrate, at the furthest, deeper than 7 millimetres (= 0·27 inch). But to thrive well, plants do not require a soil saturated with nutritive substances; for, the absorption numbers we have quoted sufficiently show how far the arable soils are from a state of complete saturation. All that plants need for their proper nutrition is that their roots, downwards in the soil, should come in contact with a certain quantity of saturated earth; and the mechanical operations of tillage have the important object of conveying earthy particles saturated with nutritive substance, and of mixing them with others, which by preceding cultivation have become poorer in those constituents.

The average crop from a hectare of wheat (2000 kilogrammes = 4400 lbs. of grain, and 5000 kilogrammes = 11,000 lbs. of straw) contains 52 million milligrammes = 114·4 lbs.) of potash, 26 million milligrammes (= 57·2 lbs.) of phosphoric acid, and 54 million milligrammes (= 118·8 lbs.) of nitrogen. Assuming the nitrogen to be supplied by the soil, the wheat plants growing on a square metre (= 10·75 square feet) receive the ten-thousandth part of the potash, phosphoric acid, and nitrogen, or altogether 13,200 milligrammes (= 203·3 grains). Supposing 100 plants to grow upon a square metre, each of these takes up from the soil 132 milligrammes of these constituents, or 54 milligrammes of nitrogen = 65 milligrammes or 1 grain of ammonia, 52 milligrammes (= 0·8 grain) of potash, and 26 milligrammes (= 0·4 grain) of phosphoric acid.

Each cubic centimetre (= ·06 cubic inch) of Bogenhausen loam absorbs to saturation 2·6 milligrammes (=·04 grain) of ammonia, 2·3 milligrammes (= 0·35 grain) of potash, and 0·5 milligrammes (= ·008 grain) of phosphoric acid; therefore, to restore a sufficiency of these constituents which the wheat plant has taken from the soil, would require a supply of 25 cubic centimetres of the saturated earth, and 25 milligrammes of phosphate of lime for each square decimetre of the field. Calculated upon a square decimetre (= 15½ square inches) of surface and a depth of 20 centimetres (=7·8 inches), these 25 cubic centimetres constitute the eightieth part of the entire mass of earth.

The experiments of Nägeli and Zoeller, before described, furnish a good example of this kind of manuring. The manure consisted of turf, partly saturated with nutritive substances and mixed with three volumes of turf almost absolutely unfruitful; this constituted a soil of the same degree of fertility as good garden mould.

Such an addition of earth saturated with mineral constituents does not usually take place; but the ordinary method of manuring comes exactly to the same result. The field is dressed with liquid or solid manuring matters containing nutritive substances, which combine immediately if in solution, gradually if requiring a certain time for solution, with the earthy particles with which they are in contact, and saturate them; and it is properly *this earth, saturated with manuring matters on its outermost surface or in the inner parts with which the farmer manures*, i. e. with which he replaces the mineral constituents withdrawn from the soil.

Experience has taught the agriculturist which parts of the soil may be enriched with nutritive substances most profitably for himself, or rather for his plants; and it is remarkable in the highest degree how he has found out the proper method of manuring in accordance with the nature of the intended crop, the soil, and the period in which the plants are developed; also whether to proceed by simple top-dressing or by ploughing the manure in to a greater or less depth.*

In these respects the successes of the agriculturist would be still greater if the nutritive substances contained in the manure principally used, namely, farm-yard manure, were more uniformly mixed and distributed, because this would lead to a more uniform distribution of them in the soil.

Farm-yard manure is a very irregular mixture of decaying straw and vegetable remains, combined with solid animal excrements, the latter constituting the smaller portion of the whole mass: it is soaked with fluids which hold ammonia and potash in solution. If a hundred samples be taken from a hundred different

* 'Journ. of the Royal Agric. Soc. England,' t. 21, p. 330.

parts of a dung-heap, the analysis of each sample will show different proportions of nutritive constituents: hence it is evident that by a dressing with farm-yard manure hardly two spots in the soil will receive the same amount of nutritive substances.

The spot occupied by a dung-heap on a field during rainy weather, will be marked in the whole period of vegetation, and often even in the second year by a more luxuriant growth of plants, especially of cereals, though the plants growing on it will not always furnish a perceptibly greater yield of grain. If the potash and ammonia received by this spot above what was required for the formation of grain, had been more evenly distributed, and thus accessible to the plants in other places, the yield of corn from those plants would have been increased; whereas the excessive accumulation in one place merely increased the yield of straw. The unequal distribution of the other ingredients of farm-yard manure in the soil leads to a similar inequality in the developement of the several parts of the cereal plants. On an ideal field, with the nutritive substances supposed to be distributed with perfect uniformity, and all accessible to the roots, all the cereal plants, other conditions being the same, should attain the same height, and each ear yield the same number and weight of grains.

In the short, rotten farm-yard manure, the nutritive substances are much more uniformly distributed than in the fresh straw manure; and the agriculturist effects a still more uniform diffusion by mixing the dung with earth, and turning it into so-called compost. *As dung and all other manuring agents act only through the medium of the earthy particles that have become saturated with the nutritive substances contained in the manure*, it is, under certain circumstances, advantageous for the farmer to prepare a saturated earth, by help of his farm-yard manure, and to use this composition, which may of course be made on the field itself. If, in accordance with Voelker's valuable experiments, we assume one cubic mètre (= 35 cubic feet) of farm-yard manure (500 kilogrammes or 1000 pounds) to contain 660 pounds of water, 6 pounds of potash, and 12 pounds of ammonia; and if this were mixed with 1 cubic mètre of earth, of which 1 cubic decimètre (= 61 cubic inches) absorbs 3000 milligrammes (= 46·2 grs.) of potash, and 6000 milligrammes (= 92·4 grs.) of ammonia; then, after the complete decay of organic matter in the manure (about 25 per cent. of its weight), and the evaporation of one-half of the water, the result would be 1¼ cubic mètre of earth fully saturated with all the nutritive substances in the manure. Soils that will absorb the stated amount of potash and ammonia are everywhere to be found, and the farmer will have no difficulty in choosing the earth most suitable for his compost heaps.

It is well known that dung exercises a mechanical action also, tending to diminish the cohesion of a compact soil, or to make a heavy soil lighter and more porous. For soils of this kind composts are not so well suited; and, instead of the earth, some very loose body ought to be substituted for mixing with the manure. Turf-dust will be found to answer the purpose best.*

If the crops obtained from many fields by manuring with farm-yard manure, bone-earth, guano, and in many cases also with wood-ashes and lime, are compared with what the same fields will yield in the unmanured state, the effect of these manures seems truly marvellous.

The yield of an unmanured field must correspond

* It is, perhaps, much more important than manuring with composts, which always involves much labour and more carriage, to take advantage of the absorbent properties of earth and turf, for fixing the nutritive substances contained in liquid manure. By covering the ground of a dunghill, on an area of 10 mètres square (=10·5 sq. feet) with a layer of loose turf, 1 mètre (= 3·3 feet) deep, a bed of 100 cubic mètres (= 3,500 cubic feet) of turf is formed, into which the liquid portion of the manure in the dunghill may safely be allowed to soak without the least risk of losing the smallest portion of its useful ingredients. The turf may then be used, like dung, for manuring, and of course must be renewed every year. On fields which are not tilled, such as meadows, liquid manure will naturally act with greater rapidity. The turf found in the neighbourhood of Munich, when reduced to powder, absorbs 7·892 grammes (=122 grains) of potash, and 4·169 grammes (= 64 grains) of oxide of ammonium, per 1000 cubic centimètres (= 61 cubic inches) weighing 330 grammes (11½ ozs.).

with the available nutritive substances which it contains; a lower crop corresponds to a smaller store of these matters. In any one of the cases stated, if we compare the amount of nutritive substances in the unmanured portion of a field with the crop which it produces, and then compare the additional nutritive substances or the quantity of dung with the increased crop, the increase appears to be beyond all proportion much greater than the additional supply. Hence we are led to suppose that the phosphoric acid, potash, and ammonia given in the manure must be much more efficacious than the substances present in the soil, or that the greater portion of them in the soil was ineffective, and that its power of production had depended chiefly upon the supply of manure. Thus it arises, that while some farmers believe that all manure can be dispensed with because tillage is enough to render a field productive, others suppose that the field can be kept fruitful only by manuring. All these views are based upon individual cases and have no general application; for neither one nor the other of the contending parties have any clear knowledge of the true causes upon which the power of production of this kind is founded.

In the experiments made in the year 1857, by order of the General Committee of the Bavarian Agricultural Union, on the action of phosphorite upon certain fields at Schleissheim deficient in phosphoric acid, the following crops of summer wheat were reaped from two plots of ground, one unmanured the other dressed, per hectare (= 2½ acres), with 241·4 kilogrammes (= 530 lbs.) of phosphoric acid, 657·4 kilogrammes (= 13 cwt.) of phosphorite decomposed by sulphuric acid:—

	1857.		
	Total crop.	Corn.	Straw.
	Kilogr. Cwt.	Kilogr. Cwt.	Kilogr. Cwt.
Manured with 657 kilogrms. of phosphate of lime....	5114·7=105·0	1301·7=25·5	3813·0=75·0
Unmanured	2301·0= 45·0	644·3=12·5	1656·7=32·5

From a chemical analysis made by Dr. Zoeller, of the Munich Laboratory, the soil of this field was found to give up to cold hydrochloric acid a quantity of phosphoric acid, which, calculated per hectare to a depth of 25 centimètres, amounted to 2376 kilogrammes = 5170 kilogrammes of phosphate of lime.

The quantity of phosphoric acid in the corn and straw of the crop reaped amounted together to:—

	kilogr.	lbs.	
From the manured plot	17·5 =	38·5	of phosphoric acid.
From the unmanured plot........	8·0 =	17·6	
Surplus obtained by manuring.....	9·5 =	20·9	

In the 657·4 kilogrammes of phosphorite the field received altogether 241·4 kilogrammes of phosphoric acid; accordingly, the surplus amounted only to $\frac{1}{25}$th of the phosphoric acid supplied in the manure.

There is nothing surprising in this result, as the additional phosphoric acid was not given to the plants but to the whole field. Had it been possible to surround each root with so much phosphoric acid or phosphate of lime as the surplus crop of corn and straw required for its formation, 9½ kilogrammes of phosphoric acid would have sufficed to double the produce of the unmanured plot; but in the way in which the manure was actually applied, every part of the field received an equal share of phosphoric acid.

Thus, of the total amount of 241·4 kilogrammes, only 9·5 kilogrammes came into contact with the roots of the plants, the remainder, though quite suitable for food, remaining inactive. To enable the plant to take up one part of phosporic acid, it was necessary to supply the field with five-and-twenty times this quantity.

On the other hand, the effect of the manure appears out of all proportion greater as compared with the store of phosphoric acid in the field.

The quantity of phosphoric acid contained in the corn and straw reaped from the unmanured plot is $\frac{1}{300}$th of the total amount of phosphoric acid in the field; that in the surplus crop is $\frac{1}{25}$th of the phosphoric acid supplied by the manure. As the manured plot gave double the produce of the unmanured, the effect of the phosphoric acid supplied by the manure is apparently twelve times greater than that of the acid originally contained in the soil.

The quantity of phosphoric acid supplied (241·4 kilogrammes) amounted to $\frac{1}{10}$th of the total quantity in the field (2376 kilogrammes). If the action of both had been alike, the surplus crop should have corresponded to the additional supply, but instead of being $\frac{1}{10}$th greater, it was double the crop obtained from the unmanured plot.

This fact is explained by the absorptive number of the Schleissheim soil for phosphoric acid or phosphate of lime.

If the store of phosphoric acid in the field had been uniformly distributed in the form of phosphate of lime (5170 kilogrammes) to a depth of 25 centimètres (9·8 inches), each cubic decimètre (61 cubic inches) would contain 2070 milligrammes (32 grains), each cubic centimètre about 2 milligrammes of phosphate of lime.

The field was manured with 657·4 kilogrammes of phosphorite in a soluble state, corresponding to 525 million milligrammes (525 kilogrammes) of pure phosphate of lime.

As determined by direct experiments, 1 cubic decimètre of Schleissheim soil absorbs 976 milligrammes of phosphate of lime. Each square decimètre received in the manure 525 milligrammes, which, dissolved by rain water in its descent through the soil, would be sufficient to saturate the earth fully, with phosphate of lime, to a depth of 5·4 centimètres (rather more than 2 inches), or to half-saturate it to a depth of 10·8 centimètres. Hence the manuring served to enrich the upper layer of the soil with phosphate of lime, not to the extent of $\frac{1}{10}$th, but to 50 per cent., and the greater part of this in a state available for the nutrition of plants. The absorptive power of the soil explains, therefore, why the crops obtained from manured fields are rather in proportion to the nutritive substances supplied in the manure, than to the store of these elements originally present in the soil.

The operation of manuring agents, severally or jointly applied, is even more marked upon soils which are poorer in nutritive substances than the field at Schleissheim above mentioned.

The following results were obtained on a field broken up for the purpose, which had not been touched by the plough for fifteen years, and had served as a pasture for sheep. The entire surface-layer of the ground at Schleissheim is 6 inches deep at most; below this there is no more soil, but a bed of rubble stones, which might be compared to a sieve with meshes an inch wide, through which the water runs freely; the crop obtained from the unmanured portion will give some idea of its sterility. Another portion was manured with superphosphate of lime; the quantity used per hectare (= 2½ acres) was 525 kilogrammes (= 10 cwt.) of phosphorite decomposed by sulphuric acid, containing 193 kilogrammes of phosphoric acid, or 420 kilogrammes (= 8 cwt.) of phosphate of lime.

Crop of winter-rye in 1858 at Schleissheim, per hectare:—

	Total crop.		Corn.		Straw.	
	Kilo.	Cwt.	Kilo.	Cwt.	Kilo.	Cwt.
Manured with phosphorite (rendered soluble by sulphuric acid) = 525·3 kilo. (10 cwt.) containing 192·8 kilo. (3·8 cwt.) $P\ O_5$, corresponding to 420 kilo. (8 cwt.) of pure phosphate of lime	1995·4 =	391·0	654·2 =	128·0	1341·2 =	200·0
Unmanured	397·6 =	7·8	115·0 =	2·3	282·6 =	5·5

Dr. Zoeller found by analysis that this field contained, per hectare, to a depth of 6 inches, only 727 kilogrammes (= 14 cwt.) of phosphoric acid.

The plot manured with phosphoric acid produced six times more corn and five times more straw than the unmanured plot. It will be observed that, however strikingly the action of manure was exhibited, this more abundant crop did not equal that in the experiment previously mentioned of the unmanured plot kept for a considerable time under culture. Upon comparing the amount of phosphoric acid contained in the two fields, we find that as the sheep pastures, to the depth of 6 inches, contained only half as much as the other (tilled but unmanured), the dressing with superphosphate was only just sufficient to make the sheep-meadow, to the depth of 8 or 10 centimètres (= 3 to 4 inches), equal to the other unmanured plot, in respect of the phosphoric acid contained in it.

These considerations explain how it is that by the absorption of nutritive substances in the upper layers of the soil a supply of these constituents or manuring ingredients, small in comparison to the total store in the ground, exercises so remarkable an action in the increase of produce, in the case of plants which draw their food chiefly from the upper layers of the arable surface soil.

If the action of the mineral constituents depends upon the sum of effective particles in certain places in the soil, the action rises with the number of particles by which the sum has been increased in these very places.

A more accurate acquaintance with the composition of arable surface soil, and its relation to the nutritive substances, together with a consideration of the nature and requirements of plants, must gradually lead to a comprehension of many other phenomena in agriculture, which hitherto are quite unexplained, and to many farmers are absolute mysteries. Although we know most accurately the general laws of the growth of plants, as far as these stand in connection with soil, air, and water, yet in many cases it is extremely difficult to discover the causes that render a soil unproductive for one culture-plant, e.g. peas, while the same soil is fruitful for other plants which require the same nutritive substances as peas, and often in still greater quantity. If the ground is rich enough in nutritive substances for these other plants, why is it that they do not act in the same way upon the peas? What causes prevent the latter from appropriating the nutritive substances, which the ground offers to other plants in a perfectly available condition? Finally, how comes it that this very soil, after a few years, will again yield a remunerative crop of peas, although by intervening harvests we have rather impoverished than enriched its store of nutritive substances; and that peas, when sown among oats, barley, or summer corn, will often yield a higher crop than when they grow alone upon a field, and have not to share with other plants the store of mineral constituents?

Analogous facts are observed in the cultivation of clover. In many districts, a field, after producing many clover crops, will become almost unfruitful for that plant.

In such cases, manuring fails in restoring to the field the power of producing clover; but after several years, during which the same field continues to give remunerative crops of cereal and tuberous plants, the soil again becomes for a while fruitful for clover.

For a considerable number of our cultivated plants we have a pretty accurate knowledge of specific manuring agents, i.e. those which have a peculiarly favourable influence upon the majority of fields. Farm-yard manure, as a rule, acts beneficially in all cases; salts of ammonia are especially valuable for cereals, superphosphate of lime for turnips; bone earth and ashes will perceptibly increase the produce of fruitful clover-fields, and, in like manner, a supply of lime will often make a field fruitful for clover, though otherwise unable to bear it.

But upon fields which have become, as it is termed, peas or clover sick, that is, have lost their power of producing these plants, all these matters otherwise favourable for their growth exercise beyond a certain time no further beneficial action. It is this fact in particular which embarrasses the practical farmer, and makes him doubt the lessons taught by science.

When the farmer is compelled to give up for many years the cultivation of plants which he had found remunerative, and science has no power to help him over his difficulties, what is the use of theory? So says the agriculturist who is himself unacquainted with the essence of theory.

It is a common error to fancy that an accurate knowledge of theory will give the power of explaining all cases that occur. Theory of itself does not explain a single phenomenon in astronomy, mechanics, physics, or chemistry; it studies and points out the causes which lie at the foundation of all phenomena, not the special causes upon which an individual phenomenon depends.

Theory requires that the causes which govern each individual case should be sought out one by one, and then the explanation is the proof or exposition of the manner in which they work together to produce the particular fact. It teaches us what to look for, and how to employ proper experiments in the discovery.

The reason why we have arrived at no conclusions about the facts just mentioned, depends chiefly upon this, that hitherto the practical farmer has troubled himself very little about the causes of those facts, as, indeed, the investigation of causes is not his proper business; while those who have undertaken this task show, by the way in which they attempt to discharge it, that they are but little acquainted with the plant as an organised being, having peculiar requirements which must be accurately known by all who would cultivate it properly.

In the following remarks I shall compare a pea-plant with a cereal, and shall call the attention of agriculturists to certain peculiarities which have to be considered in the cultivation of both plants.

A moderately moist, strong soil, not too cohesive and perfectly free from weeds, is particularly suited for peas and barley; a well-tilled, calcareous loam or marl is the best for both plants. An arable surface soil 6 inches deep suffices for barley, which, with its fine-matted roots spreading in tufts, finds a loose subsoil rather injurious than beneficial. Fresh manuring just before sowing acts powerfully on the growth of barley.

Whilst the barley-corn should not lie lower than an inch, the pea thrives best if the seed is put 2 or 3 inches deep in the soil. The roots of the pea-plant do not spread sideways but go deep into the earth; hence peas require a deep soil tilled down to the lower layers, and a loose subsoil. Fresh manure has scarcely any influence upon the growth of peas.

It results from these peculiarities of both plants, that the barley derives the conditions of its growth principally from the arable surface soil, the pea principally from the deeper layers of the soil. What the ground may contain below the depth of 6 inches is a matter of indifference for the barley; the contents of these deeper layers are everything to the pea.

If we now inquire what demands are made upon the soil by the two plants, we find from Mayer's investigations ('Results of Agricultural and Chemical Experiments, Munich, 1857,' p. 35), that the pea-seeds contain one-third more ash constituents (3.5 per cent.) than the barley-corns, and that the amount of phosphoric acid is pretty much the same in both (2·7 per cent.). Therefore, all other conditions being equal, the subsoil from which the pea derives its phosphoric acid must be as rich in that ingredient as the arable surface soil which supplies it to the barley.

The case is different with nitrogen—for the same amount of phosphoric acid, peas contain nearly twice as much nitrogen as barley. Assuming both plants to derive their nitrogen from the soil (which is, perhaps, not quite correct in the case of peas), then for every milligramme of nitrogen absorbed by the roots of the barley from the arable surface soil, twice as much must be received by the peas from the deeper layers.

These considerations throw some light, I think, upon the cultivation of peas; for this plant requires a very peculiar condition of the soil; and it is more easy to conceive that a ground exhausted by bearing peas should refuse to bear any more, than that the same soil, after the lapse of some years, should again become fruitful for this plant.

According to these considerations, and assuming an equality of the absorbent root-surface in both plants, a subsoil fruitful for peas must contain as much phosphoric acid, and twice as much nitrogen, as an arable surface soil suited for the cultivation of barley. For the phosphoric acid, the assumption is correct.

We understand, without difficulty, the beneficial effect of manure upon an exhausted barley field. Barley derives all the conditions of healthy growth from the surface soil, which is restored to its original state of productiveness by the manure applied.

But from our acquaintance with the properties peculiar to arable soil, we know that a layer 6 to 10 inches deep will retain all the ammonia potash and phosphoric acid contained in the largest quantity of manure usually applied by farmers; and this, too, so firmly that, except for some accidentally favourable circumstances, hardly a particle will ever reach the subsoil.

If a field is sown with plants which require deep ploughing, so that a sufficient portion of the rich surface is mixed with the exhausted subsoil, it is easy to understand that the latter may gradually become again fruitful for peas. The time in which this is effected depends of course upon the accidental selection of the plants grown in succession on the field.

In this view of the matter, the agriculturist has it in his power, by right management of his field, to shorten the time, and make the land again fit for successive crops of peas.

It is a fact, that many fields in the vicinity of towns will bear year after year, or every two years, abundant crops of peas, without ever becoming 'pea-sick;' and we know that the gardener, to achieve this result, has recourse to no extraordinary appliances, but merely tills his land deep and very carefully, using much more manure than the farmer can afford to do.

The frequent failure of peas is therefore not so very unaccountable; and there seems no reason why the farmer should despair of cultivating peas as often as serves his purpose, if he employ the right means to enrich his field in the proper spots with the elements of food which peas require.

In all problems of this kind, the secret of success is, not to suppose that the solution is easy, but that it is attended with great difficulties; for, if these did not exist, experimental art would long ago have found the solution.

The many unsuccessful experiments of Messrs. Lawes and Gilbert to make a clover-sick field again productive for clover, have a certain value, in as far as they show that mere experimenting leads to nothing. If I here bestow upon these experiments an attention which they do not deserve, my object is, not to submit them to a passing criticism, but to warn the practical man how he ought *not* to proceed in trying to solve his problems, if he wishes that his efforts should meet with success. The conclusions which Messrs. Lawes and Gilbert have drawn from their numerous experiments are as follows:—they found that when land is not yet clover-sick, the crop may frequently be increased by manuring with salts of potash and superphosphate of lime; that when, on the contrary, the land is clover-sick, none of the ordinary manures, whether 'artificial' or 'natural,' can be relied upon to secure a crop; and that the only way is to wait some years before repeating red clover on the same land.

It is hardly necessary to remark, that what Messrs. Lawes and Gilbert are here pleased to call conclusions, are no conclusions at all; what they have discovered has been experienced by thousands of agriculturists before them; and the only conclusion which they were permitted to draw should have been this—that in their attempts, by certain manures, to make a clover-sick field again productive for clover, they failed. In truth, they have not striven, in the remotest degree, to procure information about the causes of clover-sickness in a field, but they have simply tried different manures, in the hope of finding out one that might serve to restore the original productive power of the field, and such a manure they have not found.

Messrs. Lawes and Gilbert assume that, with respect to the soil, the clover plant bears the same relation as wheat or barley; and finding that on a field (whereon, notwithstanding the richest manure, clover had failed) an abundant barley or wheat crop was obtained the year after, it became a settled conviction with them that the failure of the clover had been caused by a specific disease generated in the soil by the cultivation of clover; this disease would attack the clover plant, but not the roots of wheat or barley.

Clover differs entirely from the cereal plants in this respect, that it sends its main roots perpendicularly downwards, when no obstacles stand in the way, to a depth which the fine fibrous roots of wheat and barley fail to reach; the principal roots of clover (as may be seen more especially with *Trifolium pratense*) branch off into creeping shoots, which again send forth fresh roots downwards.

Thus clover, like the pea-plant, derives its principal food from the layers below the arable surface soil; and the difference between the two consists mainly in this—that the clover, from its larger and more extensive root-surface, can still find a sufficiency of food in fields where peas will no longer thrive: the natural consequence is, that the subsoil is left proportionably much poorer by clover than by the pea.

Clover-seed, on account of its small size, can furnish from its own mass but few formative elements for the young plant, and requires a rich arable surface for its developement; but the plant takes comparatively little food from the surface soil. When the roots have pierced through this, the upper parts are soon covered with a corky coating, and only the fine root-fibres ramifying through the subsoil convey food to the plant.

Now, if we look at the experiments made by Messrs. Lawes and Gilbert to render a clover-sick field productive again for clover, we see, at once, that all the means employed were well adapted to enrich the uppermost layers of their field with nutritive substances for wheat and barley; but that the clover plant could derive benefit from this manuring only in the first stage of developement, while the condition of the lower layers remained unaltered, just as if the field had received no nutriment of any kind.

The manures applied by Messrs. Lawes and Gilbert were superphosphates of lime (300 lbs. of bone-earth and 225 lbs. of sulphuric acid per acre); sulphate of potash (500 lbs.); sulphate of potash and superphosphate, mixed alkaline salts (500 lbs. of sulphate of potash, 225 lbs. of sulphate of soda, 100 lbs. of sulphate of magnesia); mixed alkalis with superphosphate; further, salts of ammonia alone, and the same salts with superphosphate or mixed alkalis; farm-yard manure (15 tons), together with lime, or with lime and superphosphate, or with lime and alkalis in the most varied proportions; then soot; soot with lime; soot with lime, alkalis, and superphosphate. None of these manures had the slightest effect; the clover-sick field continued just as unproductive for clover as before.

The reason why these manures were inoperative is not difficult to find. Messrs. Lawes and Gilbert, in their report, leave us, indeed, in the dark as to the nature and condition of the soil upon which their experiments were made; but from some incidental observations in previous papers, we know that the fields at Rothamstead consist of a rather heavy loam, very well suited for cereals, and especially for barley.

From experiments upon the absorptive power of loam, we may assume, without risk of error, that one cubic decimètre (=61 cubic inches of loam) will absorb 2000 milligrammes (=31 grains of potash), and 1000 milligrammes (=15·5 of phosphate of lime).

The surface of an acre of loam (=405,000 square decimètres) will therefore absorb to a depth of 1 decimètre (=4 inches) 805 kilogrammes (=1,771 lbs.) of potash, and 405 kilogrammes (=891 lbs.) of phosphate of lime.

The most copious dressing with sulphate of potash which Messrs. Lawes and Gilbert gave to their field amounted to 500 lbs. (=270 lbs.) of potash; the most copious of the superphospate dressings represented 300 pounds of phosphate of lime.

Had Messrs. Lawes and Gilbert put upon the field the sulphate of potash and the phosphate of lime in a state of complete solution, the whole quantity of potash employed would have penetrated no deeper than 2 centimètres, or not quite an inch, and the phosphate of lime no deeper than 4 centimètres, or a little more than 1·6 inch. Both manures, however, were strewed over the field and ploughed in; still it cannot be assumed that the layers below a depth of 8 inches could have received any considerable quantity of potash or phosphate of lime.

At page 10 of their paper ('Report of experiments on the growth of red clover by different manures') Messrs. Lawes and Gilbert say, 'Those who have paid attention to the spread of disease in clover, on land which is said to be clover-sick, will have observed, that however luxuriant the plant may be in the autumn and winter, it will show signs of failure in March or April.' The same fact was observed in all their experiments. A field on which clover had failed was sown with barley, and when this had yielded a rich crop, another attempt was made with clover.

'The plants (say Messrs. Lawes and Gilbert) stood tolerably well during the winter, but as the spring advanced they died off rapidly.' There cannot be the slightest doubt about the reason of this decay; the exhausted subsoil had not received back any of the lost conditions of fertility, and thus the plants were starved as soon as they had pushed through the arable surface soil, and their roots were beginning to spread in the subsoil.

If the failure of the clover was attributable to a disease, this must have been of a very singular nature, as the richly-manured arable soil showed no traces of it, and it was only the subsoil which was clover-sick. The notion that there is any disease engendered by the cultivation of clover is refuted most completely, though unconsciously, by Messrs. Lawes and Gilbert themselves. They say, page 17, 'Before we enter upon the probable causes of the failure in clover, it may be well to give the results of some experiments conducted in the kitchen-garden at Rothamstead. The soil was in ordinary garden cultivation, and has probably been so for two or three centuries. Early in 1854, the $\frac{1}{500}$th of an acre (about 9¾ square yards) was measured off and sown with red clover on March 29. From that time to the end of 1859 fourteen cuttings have been taken without any resowing of seed. In 1856 this little plot was divided into three equal portions, of which one was manured with gypsum, another with sulphates of potash, soda, and magnesia, and superphosphate of lime.'

'The estimated total amount of green clover obtaied from this garden soil in six years, without further manure, is about 126 tons per acre, equal to about 26½ tons of hay. In four years the increase by the use of gypsum amounted to 15½ tons of green clover. The increase in the four years by the use of the alkalis and phosphate is estimated to amount to 28¾ tons of green produce.'

'It is worthy of remark,' continues the report, 'that it was in some of the very same seasons in which these heavy crops of clover were obtained from the garden soil, that we entirely failed to get anything like a moderate crop of clover in the experimental field, only a few hundred yards distant.'

It is, indeed, most worthy of remark, that upon the experimental field the earth was poisoned by the vegetation of the clover, so as to render it incapable of further bearing this plant; while, at the very same time, under like climatic conditions, the self-same clover-plant engendered no poison in the rich garden soil.

A comparative examination of the garden and of the field-soil seems never to have been thought of, since the two agricultural chemists were, as we before remarked, in search of an efficient manure, not of the cause of the failure of the plant. But though they have not found the smallest shred of a fact which might serve in any way to explain the strange behaviour of the clover-plant upon the two fields, they do not hesitate to present the farmer with the following ingenious explanation :—

'Among plants,' say they, 'there are certain kinds which are peculiarly circumstanced with respect to the nature of their food; the cereals, among others, feed principally upon inorganic matters, whilst others, the leguminous plants, e. g. clover, are dependent for luxuriant growth, more or less, upon a supply within the soil of complex organic compounds.'

Taking their stand upon the fact that they have failed to discover any explanation, which, in their opinion, they surely must have done, had it been possible to find one, they coolly ask us to believe that there are, among the higher classes of plants, certain species bearing about the same relation to other species as the carnivorous to the graminivorous animals; and as the former feed upon complex organic compounds prepared in the bodies of the latter, so it is, also, with the clover-plant; like mushrooms, it represents the carnivorous order in the vegetable kingdom.

It is hardly worth while to take any notice of this explanation; but it might still prove useful to inquire whether, apart from all consideration of the absorptive power of the soil, Messrs. Lawes and Gilbert have really exhausted all the means that might have been employed to restore the productiveness of the clover-sick field for clover, so as to be justified in giving it as their opinion that when land is clover-sick, none of the ordinary manures, artificial or natural, can be relied upon to secure a crop.

We may ask why Messrs. Lawes and Gilbert did not, instead of superphosphate of lime, try bone ash, the action of which extends much deeper than that of the superphosphate; and why sulphate of potash and sulphates alone were employed? It is not impossible that common wood ashes might have proved more effective than sulphate of potash; and, above all, chloride of potassium ought to have been tried, which, as an ingredient of liquid manure, is more useful to clover than any other of the potash salts. It is also difficult to understand why liquid manure was not employed, and why chloride of sodium was excluded from the list of manuring agents. If we consider what Messrs. Lawes and Gilbert omitted to do in their endeavour to solve the problem, and what they ought to have done, the conclusion is inevitable, that they had no accurate notion of the nature of their task.

Now, the want of a proper insight into the nature of a phenomenon which is to be investigated is surely the greatest of all difficulties in the way of attaining a practical result. If the unproductiveness of a field for clover and peas depends upon a want of nitrogenous food in the deeper layers of the soil, and upon no other cause, the absorptive power of the various soils for ammonia renders it extremely difficult to enrich the subsoil with this element of food. But the case is quite different with the nitrates, which penetrate to any depth, as the nitric acid is not absorbed by the soil; probably, nitrate of soda may afford a means of making a field productive for clover or peas, in cases where there is a deficiency of nitrogenous food.

As manuring with burnt lime is often found beneficial for clover and also for peas, and a calcareous soil tends, in a special degree, to promote the formation of nitric acid, it is not improbable that it is owing to this property that lime promotes the growth of deep-rooting plants by converting ammonia into nitric acid, and causing nitrogenous food to find its way to the deeper layers of the soil.

CHAPTER IV.

FARM-YARD MANURE.

The fertility of a soil depends upon the sum of available food, the continuance of the fertility upon the total amount of all food in it—Chemical and agricultural exhaustion of the soil—Exhaustion of the soil by cultivation, laws regulating its progression; effect of the transformation in the soil of the chemically fixed into physically fixed elements of food; effect on the progress of exhaustion by partial restoration of the withdrawn food of plants—Progress of the exhaustion by different cultivated plants—Cultivation of cereals, consequence of removing the grain and leaving the straw in the soil; intervening clover and potato crops; effect of leaving in the ground the whole or a portion of these crops; division of soils; productive power of wheat fields increased by accumulating in them the materials derived from clover and potato fields; cultivation of fodder plants; their food partly derived from the subsoil; addition of these increases the productive power of the surface soil—Natural connection between the cultivation of cereals and fodder plants, the influence on the fertility of land—Exhaustion of the soil removed by the restoration of the withdrawn mineral constituents; the excrement of men and animals contains these; their restoration depends upon the agriculturist.

TO form a correct idea of the effects produced by farm-yard manure in husbandry, it must be remembered that the fertility of a soil is always exactly proportionate to the amount which it contains of nutritive substances in a state of physical combination; and that the permanence of its fertility or its productive power stands in proportion to the total quantity of the constituents in the soil capable of passing over into that physical condition.

The amount of crop reaped from a field in a given time is proportionate to that fraction of the total constituents which has passed during this time from the ground into the plants grown upon it. If one of two fields yields twice as large a crop of wheat and straw as the other, this necessarily presupposes that the wheat-plants upon the one field have received from the ground twice as much nutriment as those upon the other.

If the same or different plants are cultivated in succession on a field, the crops will gradually decrease, and the soil will be termed 'exhausted,' in an agricultural sense, when the crops cease to be *remunerative*, i e. do not cover the expense of labour, interest of money, &c. As the high crops were caused by the soil giving to the plant a certain number of parts from the total nutritive substances, just so the exhaustion of the field proceeds from a diminution in the sum of those nutritive substances.

The same number of plants cannot thrive upon the same field as formerly, if the same quantity of nutritive substances enjoyed by the previous crop is no longer to be found. The exhaustion of a cultivated field in a chemical sense differs from the agricultural use of the term in this, that the former refers to the total amount of nutritive substances in the soil, the latter to that portion only of the total amount which the ground can furnish to plants. A field is termed exhausted in a chemical sense when it altogether fails to produce any more crops.

Of two fields, one of which contains, to the same depth, a hundred times, the other only thirty times, the amount of food required by a full wheat crop, the former furnishes to the roots of the plants more nutriment than the latter in the proportion of 10 : 3, supposing the condition and mixture of the soil to be the same in both cases. If the roots of a plant receive from certain spots of the one field 10 parts by weight of nutriment, the roots of the same plant will find upon the other field only 3 such parts available for absorption.

An average wheat crop of 2000 kilogrammes (= 39 cwts.) of grain, and 5000 kilogrammes (= 98 cwts.) of straw, receives from a hectare (= 2½ acres) of ground 250 kilogrammes (= 5 cwts.) of ash-constituents on, an average. Now, upon the supposition that a field, to give an average crop, must contain 100 times that quantity (or 25,000 kilogrammes) of ash-constituents in a perfectly available state, it follows that such a field gives 1 per cent. of its total store to the first crop.

The soil will still continue productive for new wheat crops in the following years; but the amount of produce will gradually decrease.

If the soil is most carefully mixed, the wheat plants will, in the next year, find everywhere upon the same field 1 per cent. less nutriment, and the produce in corn and straw must be smaller in the same proportion. If the climatic conditions, the temperature, and the fall of rain remain the same, there will be reaped from the field in the second year only 1980 kilogrammes of grain, and 4950 kilogrammes of straw; and in each succeeding year the crop must fall off in a fixed ratio.

If the wheat crop in the first year took away 250 kilogrammes of ash-constituents, and the soil contained per hectare to the depth of 12 inches one hundred times that quantity (25,000 kilogrammes), there would remain in the ground at the end of the thirtieth year of cultivation 18,492 kilogrammes of nutritive substances.

Whatever variations in the amount of produce may have been caused by climatic conditions during the intervening years, it is evident that in the thirty-first year, if there has been no restoration of mineral matters, the field will produce, even under the most favourable circumstances, only $\frac{185}{250} = 0{\cdot}74$, or somewhat less than three-fourths of an average crop.

If these three-fourths of an average crop do not give the farmer a sufficient excess of income over expenditure, if they barely cover his outlay, the crop can no longer be called remunerative. He calls his field 'exhausted' for the cultivation of wheat, although it contains seventy-four times the quantity of nutritive substances required by an average crop for the year. Owing to the presence of the entire sum of nutritive substances, in the first year of cultivation each root found, in the parts of the soil in contact with it, the requisite amount of mineral food for its complete develepement; but, owing to the continuous crops, only three-fourths of this quantity is found in the thirty-first year in the same portions of the soil.

An average crop of rye (1600 kilogrammes ($= 31\frac{1}{3}$

cwts.) of grain, and 3800 kilogrammes (= 74½ cwts.) of straw) takes away from the ground per hectare only 180 kilogrammes (= 3½ cwt.) of ash-constituents.

If the production of an average wheat crop requires the presence in the soil of 25,000 kilogrammes of the ash-constituents of wheat plants, a soil with only 18,000 kilogrammes of such constituents will prove sufficiently rich to give an average and a succession of remunerative crops of rye.

By our reckoning, a field, though exhausted for the cultivation of wheat, still contains 18,492 kilogrammes of mineral constituents, the same in properties as those which the rye plant requires.

If it is asked after how many years continuous rye-cultivation the average crop will sink down to a three-quarter crop, assuming this to be no longer remunerative, we find that the field will produce 28 remunerative rye-crops, and after 28 years will be exhausted for its cultivation.

The nutritive substances yet remaining in the soil will still amount to 13,869 kilogrammes of ash-constituents.

A field on which rye can no longer be cultivated with profit is not on that account unfruitful for oats.

An average crop of oats (2000 kilogrammes (= 39 cwts.) of grain, and 3000 kilogrammes (= 59 cwts.) of straw) takes from the soil 310 kilogrammes (= 6 cwts.) of ash-constituents, being 60 kilogrammes (= 1·2 cwt.) more than is removed by a wheat crop, and 130 kilogrammes (= 2½ cwts.) more than by a rye crop. If the absorbent root-surface of the oat plant were the same as that of rye, oats after rye would not yield a remunerative harvest; for a soil supplying, for the production of a crop of oats, 310 kilogrammes out of a stock of 13,869 kilogrammes, loses thereby 2·23 per cent. of its store of mineral constituents, whereas the roots of rye extract only 1 per cent.

To produce a remunerative crop of oats after rye is only possible when the root-surface of the oat plant exceeds that of the rye in the proportion of 2·23 to 1.

Oat crops will therefore exhaust the soil the most speedily; after 12¾ years the harvest will sink to three-fourths of the original amount.

None of the causes tending to diminish or increase the crops have any influence on this law of exhaustion of the soil by cultivation. Whenever the stock of nutriment has been lowered to a certain point, the ground ceases to be productive, in an agricultural sense, for cultivated plants.

For every cultivated plant such a law exists. This state of exhaustion will inevitably take place, even though only a single one of the various mineral constituents required for the nutrition of the plants has been withdrawn from the soil by a succession of crops; for the one constituent which fails or is deficient renders all the rest ineffective. With each crop, each plant, or portion of a plant, taken away from a field, the soil loses part of the conditions of its fertility, that is, after a course of years of cultivation it loses the power of again producing this crop, plant, or part of a plant. A thousand grains of corn require from the soil a thousand times as much phosphoric acid as one grain; and a thousand straws demand a thousand times as much silicic acid as one straw. When, therefore, the soil is deficient in the thousandth part of phosphoric or silicic acid, the thousandth grain or the thousandth straw will not be formed. If a single stalk of corn is taken away from a field, the consequence is that the field no longer produces one straw in its room.

Hence it follows that a hectare of ground, containing 25,000 kilogrammes of the ash-constituents of wheat, uniformly distributed, and presented to the roots of the plants in a perfectly available condition, can, up to a certain point, continue to give in succession remunerative crops of various cereal plants, without receiving any restoration of the mineral constituents taken away in the corn and straw, provided that the uniform mixture of the soil be maintained by careful ploughing and other suitable means. The succession of crops is determined by this principle, that the second plant must always take away from the soil less than the first, or possess a greater number of roots, or generally a larger absorbent root-surface. After the average crop of the first year, the crops would go on yearly diminishing.

The farmer, to whom uniform average harvests are the exception, and an alternation of good and bad crops dependent upon change of weather is the rule, would hardly notice this constant diminution, even supposing his field to be actually in that favourable chemical and physical condition which would enable him to cultivate wheat, rye, and oats for seventy years in succession, without restoring any of the mineral constituents removed from the soil. Good crops approaching the average in favourable years, would alternate with deficient crops in bad seasons; but the proportion of unfavourable to favourable returns would go on increasing.

Most of the land under cultivation in Europe is not in the physical condition assumed in the case of the field which we have been considering.

In most fields the phosphoric acid required by the plants is not all distributed in an effective condition, and accessible to the roots; a part of it is merely disseminated through the soil in the form of small granules of apatite (phosphate of lime); and even where the soil contains altogether a quantity more than sufficient, yet in some parts of it there is much more and in others less than the plants require.

If we suppose our field to contain 25,000 kilogrammes of the ash-constituents of wheat equally distributed through the soil, and five, ten, or more thousand pounds of the same constituents, unequally distributed, the phosphoric acid as apatite, the silicic acid and potash as decomposable silicates; and, further, if every two years a certain quantity of this second portion of food elements becomes, in the manner stated, soluble and distributable, so that the roots of plants in all parts of the arable soil could find as much of these nutritive substances as in the preceding years of cultivation—sufficient, therefore, for an average crop; we should, in that case, be able to obtain full average crops for a number of years by always letting a year of fallow intervene after a year of cultivation. Instead of thirty progressively decreasing crops, we should in that case reap thirty full average crops in sixty years, if the excess of mineral matter in the soil were sufficiently large to replace everywhere the phosphoric acid, silicic acid, and potash taken away in each year of crops. After the exhaustion of this excess of mineral matter, the period of diminishing crops would commence for our field, and the interposition of fallow years would, after this, no longer exercise the least influence on the production of larger crops.

If the excess of phosphoric acid, silicic acid, and potash, which we have assumed in the case under consideration, were not unequally but uniformly distributed, and everywhere perfectly accessible and available to the roots of the plants, our field would be able to yield thirty full average crops in thirty successive years, without the intervention of a season of fallow.

Let us return to our field, which we have assumed to contain 25,000 kilogrammes of the ash-constituents of wheat, equally distributed through the soil, and in a suitable state for absorption by the roots. Suppose we were to cultivate wheat upon it year after year, but instead of removing the entire crop we were merely to cut off the ears, leaving the straw on the ground and immediately ploughing it in; the loss sustained by the field would, in this case, be less than before, as all the constituents of the straw and the leaves would be left in the field, the mineral constituents of the grain alone having been removed.

The straw and leaves contain, among their constituent elements, the same mineral substances as the grain, only in different proportions. If the total quantity of phosphoric acid conveyed away in the straw and corn be designated by the number 3, the loss will be only 2, if the straw is left in the ground. The decrease of produce from the field, in the following year, is always in proportion to the loss of mineral substances occasioned by the preceding crop. The next produce of grain will be a little larger than it would have been had the straw not been left in the ground; the produce of straw will be nearly the same as in the preceding year, because the conditions for the formation of straw have been but slightly altered.

Thus, then, by taking away from the ground less than formerly, we increase the number of remunerative crops, or the sum total of grain produced in the whole series of corn harvests. Some of the straw-constituents are converted into corn-constituents, and are now removed from the field in the latter form. The period of final exhaustion, though sure to come in the end, will, under these circumstances, occur later. The conditions for the production of grain go on continually decreasing, because the substances removed in the corn are not replaced.

It would make no difference in this respect, if the straw were cut and carted about the field, or used as litter for cattle and then ploughed in; the supply thus bestowed upon the field, having been originally taken from the field, cannot enrich it.

Considering that the combustible elements of the straw are not supplied by the soil, it is clear that in leaving the straw in the ground we leave nothing more than the ash-constituents of the straw. The field remained somewhat more fruitful than before, because a little less had been taken away.

If the corn or its ash-constituents were ploughed in with the straw, or if, instead of it, a corresponding quantity of some other seed containing the same ash-constituents as wheat, e. g. ground rape-cake, that is, rape-seed freed from the fatty oil, were given in proper proportion to the ground, its composition would remain the same as before: the next year's crop would equal that of the preceding year. If after every harvest the straw is always in this manner returned to the field, the further consequence will be an inequality in the composition of the effective constituents in the arable soil.

We have supposed our field to contain the ash-constituents of the entire wheat plant in proper proportion for the formation of straw, leaves, and grain. By leaving the straw-constituents in the ground while continually removing the grain-constituents, the former will accumulate and grow out of due proportion to the remainder of the grain-constituents still contained in the field. The field retains its fertility for straw, but the conditions required for the production of grain are diminished.

The consequence of this disproportion is an unequal developement of the entire plant. As long as the soil contained and supplied the right proportion of ash-constituents needful for the uniform growth of all parts of the plant, so long the quality of the seed and the ratio between straw and corn in the diminishing crops remained constant and unaltered. But, in proportion as the conditions for the production of leaves and straw became more favourable, the quality of the grain deteriorated with its decreasing quantity. The distinctive mark of this inequality in the soil, resulting from cultivation, is a decrease in the weight of the bushel of corn reaped from the field. At first a certain quantity of the constituents restored to the soil in the straw (phosphoric acid, potash, magnesia), was expended in the formation of grain; but afterwards the case is reversed, and the grain-constituents (phosphoric acid, potash, magnesia) are drawn upon for the production of straw. The condition of a field is conceivable where by reason of inequality in the relative conditions for producing straw and grain, under temperature and moisture favourable for the formation of leaves, a cereal plant may yield an enormous crop of straw, with empty ears.

The farmer, in cultivating his plants, can act upon the direction of the vegetative force only through the soil, i. e. by supplying his field with nutritive substances, in the right proportions. For the production of the largest crop of grain, the soil must contain a preponderating quantity of the nutritive substances necessary for the formation of seed. For leafy plants, turnips, and tuberous plants, the proportion is reversed.

It is therefore evident, that if on our field containing 25,000 kilogrammes of the ash-constituents of the wheat-plant, we cultivate potatoes and clover, and take away from the field the entire crop of tubers and clover, we remove from the ground, in these two products, as much phosphoric acid and three times as much potash as in three wheat crops. It is certain that the abstraction of these important mineral constituents from the ground, by the cultivation of another plant, must greatly affect the fertility of the soil for wheat; the crops of wheat diminish in amount and in number.

But if, instead of this, we were to cultivate on our field alternately, wheat one year, potatoes the next, leaving the entire potato crop, tubers included, and the wheat straw on the ground to be ploughed in, and if this alternation of crops were continued for sixty years, the crop of corn which the field was originally capable of yielding would not in the slightest degree be altered or increased. The field would gain nothing by the cultivation of potatoes; and would lose nothing, because the whole crop was left in the soil. When by taking corn crops from the field, the store of mineral constituents had been reduced to three-fourths of the original quantity, the field would cease to furnish remunerative crops, supposing that three-fourths of an average harvest leave no margin of profit for the farmer. The same results would follow, if instead of potatoes we interpose clover, and constantly ploughed it in. We have assumed the field to be in the best physical condition, which therefore could not be improved by the incorporation of the organic substances of the clover and the potatoes. Even if we were to take the potatoes from the field, to mow down and dry the clover, giving both to cattle in the farm-yard or making any other use of them, and then to bring all back to the field and plough them in, so as to restore to the soil all the mineral constituents contained in both crops, yet by all these operations the field would not produce, in thirty, sixty, or seventy years, a single grain of corn more than without this alternation. The conditions required for the production of grain are not improved in the field during the whole of this period, and the causes of decrease in the crops remain the same.

The ploughing in of the potatoes and the clover could have a beneficial effect upon those fields only which have an inferior physical condition, or in which the mineral constituents are unequally distributed, or are partially inaccessible to the roots of plants. But this effect is like that of green manuring, or of one or more years of fallow.

By the incorporation of the clover and the organic constituents with the soil, its store of decaying substances and nitrogen increased year by year. All that these plants received from the atmosphere remained in the ground; but the increase of these otherwise so useful substances cannot make the soil produce a larger amount of grain than before; since the production of grain depends upon the right proportion of ash-constituents in the soil, and these, so far from being increased, have been gradually reduced by the removal of the corn crops. The augmentation of nitrogen and of decaying organic substances in the soil might possibly lead to an increase of produce for a number of years; but the period when this field will cease to give remunerative crops will in that case come all the sooner.

If we take three wheat fields, and cultivate wheat upon the one, potatoes and clover upon the other two; and suppose we remove the corn alone from the wheat field and heap upon it and plough in all the crop of clover and all the potato tubers, then the wheat field will be more fertile than before, for it has been enriched by all the mineral constituents which the two other fields had furnished to the potatoes and the clover. It has received three times as much phosphoric acid and twenty times as much potash as was contained in the corn crop it produced.

This wheat field will now be able to produce three full corn crops in three successive years, because the conditions for the formation of straw have remained unaltered, while those for the production of grain have been increased three-fold. If the farmer by this method raises as much corn in three years as he could obtain from the same fields in five years without the addition and cooperation of the constituents contained in the clover and the potatoes, it is clear that his profit has been greater, since with three seed-corns he has obtained as good a harvest as in the other case with five. But what the wheat field has gained in fertility, the other two fields have lost; and the final result is, that at less cost of cultivation, and with more profit than before, his three fields are brought to the period of exhaustion which inevitably results from the continued removal of the mineral constituents in the crops of corn.

The last case which we have to consider is when the farmer, instead of growing potatoes and clover, cultivates turnips and lucerne, which by their long penetrating roots extract a great quantity of mineral constituents from the subsoil, to which the roots of the cereals very seldom penetrate. When the fields have a subsoil favourable to the growth of these plants, it is as though the arable surface soil were doubled. If the roots of these plants receive the half of their mineral nutriment from the subsoil, and the other half from the arable surface soil, the latter will lose by these crops only half as much as they would, if all the mineral constituents had been drawn by them from the surface.

Thus the subsoil, considered as a field apart from the arable soil, gives to turnips and lucerne a certain quantity of mineral constituents. Now, if the whole of the turnip and lucerne crops were ploughed in during the autumn in a wheat field which had yielded an average crop of wheat, so that the field should receive back more than it had lost in the corn, it is clear that this field might be maintained in an equable state of fertility, at the expense of the subsoil, just so long as the latter remained productive for turnips and lucerne.

As, however, turnips and lucerne require for their developement a very great quantity of mineral constituents, the subsoil is so much the sooner exhausted, when it contains fewer of such constituents. Now as it is not actually severed from the arable surface, but lies underneath, it can scarcely regain any of all the constituents which it has lost, because the surface soil intercepts and retains the portion supplied. Only that part of the potash, ammonia, phosphoric acid, and silicic acid, which is not taken up and fixed by the surface soil, can reach the subsoil.

It is therefore possible, by the cultivation of these deep-rooting plants, to gain an abundant supply of nutritive substances for all plants drawing their nutriment chiefly from the arable soil; but this supply is not lasting, and in a comparatively short time many fields will cease to bear crops, because the subsoil is exhausted, and its fertility is not easily restored.

If a farmer grows upon three fields, potatoes, corn, and vetches or clover, alternately, or if he cultivates one field with potatoes, corn, and vetches successively, selling the crops, and going on in the same way for many years, without manuring, any one can foresee the end of such husbandry, because such a system cannot possibly last. No matter what plants may be selected, what variety of cereals, tuberous or other plants, or in what rotation, the field will at length be reduced to such a state that the cereals will yield no more than the seed sown, the potatoes will give no tubers, and the vetches or clover will die away after barely appearing above ground.

From these facts it follows indisputably, that there is no plant which spares the ground, and none which enriches it. The practical farmer is taught by innumerable instances that the success of a second crop depends upon the previous one, and that it is by no means a matter of indifference, in what order he cultivates his plants; by previously cultivating some plant with extensive ramification of roots, the soil is made fitter for the growth of a succeeding cereal, which will now thrive better, even without the application of manure (with sparing application), and yield a richer crop. But this is not a saving of manure for future crops, nor has the field been enriched in the conditions of its fertility. There has been an increase, not in the sum of the nutriment, but in the available particles of that sum, and their operation has been hastened in point of time.

The physical and chemical condition of the field was improved; but the store of chemical elements was reduced. All plants, without exception, drain the soil, each in its own way, and exhaust the conditions for their reproduction.

In the produce of his field the farmer actually sells his land; he sells, in his crops, certain elements of the atmosphere, which come of themselves to his soil; and with them certain constituents of the ground, which are his property, and which have served to form, out of the atmospheric elements, the body of the plant, being themselves component parts of that body. In alienating the crops of his field, he robs the land of the conditions required for their reproduction. Such a system of husbandry may properly be called a system of spoliation.

The constituents of the soil are the farmer's capital; the atmospheric nutritive substances are the interest of his capital; with the former he produces the latter. In selling the produce, he alienates part of his capital and the interest; in restoring the constituents of the soil to the ground, he retains his capital.

Common sense tells us, and all farmers agree, that clover, turnips, hay, &c., cannot be sold off from a farm without materially damaging the productive power of the land for corn.

Everyone willingly admits, that the removal of clover is prejudicial to the cultivation of corn; but that the removal of corn should injure the cultivation of clover is to most farmers an inconceivable, nay, an impossible idea.

Yet the natural connection and mutual relations

between the two classes of plants are as clear as daylight. The ash-constituents of clover and corn are the conditions for the formation of clover and corn, and are identical as far as the elements are concerned.

Clover, just like corn, requires for its production a certain amount of phosphoric acid, potash, lime, and magnesia. The mineral constituents of clover are the same as those of corn, *plus* a certain excess of potash, lime, and sulphuric acid. The clover draws these constituents from the soil, the cereal plants may be represented as deriving them from the clover. In selling his clover, therefore, the farmer takes away the conditions for the production of corn, and there remains behind in the soil less nutriment for the corn; if he sells his corn, he takes away from the land some of the most indispensable conditions for the production of clover, hence the clover crop fails in the subsequent year.

The peasant knows the operation of these fodder-plants, and expresses his views in his own way when he says, 'that, as a matter of course, a man must not sell his manure, without which no permanent cultivation is possible, and that in selling the fodder-plants, a man sells his manure.' But that in selling his corn, a farmer is still parting with his manure, does not seem to be understood by many even of the most enlightened agriculturists. Farm-yard manure contains all the mineral constituents of fodder; and these consist of the constituents of corn, *plus* a certain quantity of potash, lime, and sulphuric acid. It is quite evident, that as the whole dung-heap consists of parts, not one of those parts should be alienated; and if it were possible, by any means, to separate the corn-constituents from the rest, they would possess the greatest value to the farmer, because upon them the cultivation of the corn depends. But this separation actually takes place in the growth of corn, as the mineral constituents of the manure become the constituents of the corn; hence in selling the corn, the farmer alienates a portion, and indeed the most efficient portion, of his manure.

Two dung-heaps, looking quite alike, and apparently

of the same quality, may yet have a very dissimilar value for the cultivation of corn. If in one heap the ash-constituents of corn are twice as many as in the other, the former has double the value of the other. By the removal of the mineral constituents of the corn, which were derived from the manure, the efficacy of the manure with regard to future corn crops is constantly diminished.

From whatever point of view, therefore, the alienation of corn or other field produce may be regarded, the farmer who does not replace the mineral constituents taken away in the crops, will find that the inevitable result is exhaustion of the soil. Continued removal of the corn crops makes the ground unproductive for clover, or deprives the manure of its efficacy.

In our exhausted fields the roots of cereals no longer find, in the upper layers of the soil, sufficient nutriment for the production of a full crop: the farmer, therefore, grows on these fields clover, turnips, and other plants of the kind, which, with their wide-spreading and deep roots, penetrate in all directions through the soil, open up the ground by their large root-surface, and appropriate the constituents which are needed by cereals for the formation of seed. In the residue of these plants, in the constituents of the stalks, the roots and the tubes, which the farmer puts upon the arable surface in the form of manure, he restores to the land, in a concentrated form, the corn-constituents for one or several full crops: what was below and scattered, is now above. The clover and the fodder-plants did not engender the conditions of richer corn-crops, any more than rag-gatherers produce the conditions for paper-making: they are mere collectors.

From the foregoing remarks it is evident that the cultivation of plants exhausts the fertile soil, and renders it unfruitful. In selling the produce of his fields, which serves as food for man and beast, the farmer removes a portion of his soil, and indeed the constituents most efficient for the production of future crops. In course of time, the fertility of his fields will decrease, no matter what plants he cultivates, or what order of rotation he may adopt. The removal of his crops is nothing else than robbing the ground of the conditions for future harvests.

A field is not exhausted for corn, clover, tobacco, or turnips, so long as it yields remunerative crops, without needing the replacement of those mineral constituents which have been carried away. It is exhausted from the time that the hand of man is needed to restore the failing conditions of its fertility. In this sense most of our cultivated fields are exhausted.

The life of men, animals, and plants is most intimately connected with the restoration of all those conditions which cause the vital process to go on. The soil, by its constituents, takes part in the life of the plant; its permanent fertility is inconceivable and impossible, without the replacement of those conditions which have made it productive.

The mightiest river which sets in motion thousands of mills and machines must fail, if the streams and brooks supplying its waters run dry; so, too, the streams and brooks will run dry if the many little drops of which they consist fail to return in the form of rain to the place whence their sources spring.

A field which, by the successive cultivation of different plants, has lost its fertility, may recover the power of yielding a new series of crops of the same plants, by the application of manure.

What is manure, and whence comes it? All manure comes from the farmer's fields: it consists of straw, which has served as litter; of remains of plants, of the liquid and solid excrements of men and animals. The excrements are derived from food.

In his daily bread, man consumes the ash-constituents of the grain from the flour of which bread is made: in meat he consumes the ash-constituents of flesh.

The flesh of herbivorous animals, and its ash-constituents, are derived from plants; these ash-constituents are identical with those of the seeds in leguminous plants. Hence if an entire animal is burnt to ashes, the residue will differ little from the ashes of beans, lentils, and peas.

In bread and flesh, therefore, man consumes the ash-constituents of seed, or of seed-constituents which the farmer has obtained from his fields in the form of flesh.

Of the large amount of mineral substances which man consumes in his food during a lifetime, but a small fraction remains in his body. The body of an adult does not increase in weight from day to day, which proves that all the constituents of his food must completely pass out again from his system.

Chemical analysis demonstrates that the excrements of man contain the ash-constituents of bread and flesh very nearly in the same quantity as they exist in the food, which in the body undergoes a change similar to that which would take place in a furnace.

The urine contains the soluble, the solid excrements the insoluble ash-constituents of food: the stinking substances are the smoke and soot of an imperfect combustion. With these are mixed up the undigested and the indigestible remains of food.

The dung of swine fed on potatoes contains the ash-constituents of the potato; that of the horse, the ash-constituents of hay and oats; that of cattle, the ashes of turnips, clover, and other plants which have served them as food. Farm-yard manure comprises a mixture of all these excrements.

That farm-yard manure will completely restore the fertility of a field exhausted by cultivation is a fact fully established by the experience of a thousand years.

Farm-yard manure supplies to the field a certain quantity of organic, i. e. combustible substances, together with the ash-constituents of the food consumed. We must now consider what part is taken, in the restoration of fertility, by the combustible and incombustible constituents of the manure.

The most superficial examination of a cultivated field shows that all the combustible constituents of the plants grown upon it are derived from the air and not from the soil. If the carbon even of a portion of the vegetable matter in the crop were derived from the soil, it is quite clear, that if the ground contained a certain amount of carbon before the harvest, this amount must be smaller after every harvest. A soil deficient in organic matter must necessarily be less productive than a soil abounding in it.

Now, experience proves that a field in constant cultivation does not, therefore, become poorer in organic or combustible substances. The soil of a meadow which in ten years has yielded a thousand cwt. of hay per hectare, is found to be, at the end of those ten years, not poorer in organic substances, but richer than before. A clover-field after a crop retains in the roots left in the ground more organic substances, more nitrogen, than it originally possessed; yet after a number of years it becomes unproductive for clover, and no longer gives remunerative returns of that crop.

A field of wheat, or potatoes, is not poorer in organic substances after harvest, than before. As a general rule, cultivation increases the store of combustible constituents in the ground, while its fertility, however, steadily diminishes. After a consecutive series of remunerative crops of corn, turnips, and clover, these plants will thrive no longer in the same field.

Since, then, the presence of decaying organic remains in the soil does not, in the slightest degree, prevent or arrest its exhaustion by cultivation; it is impossible that an increase of those substances can restore the lost capacity of a field for production. In fact, when a field is completely exhausted, neither boiled saw-dust nor salts of ammonia, nor both combined, will impart the power of yielding the same series of crops a second and third time. When these substances improve the physical condition of the ground, they exert a favourable influence upon the produce; but, after all, their ultimate effect is to accelerate and complete the exhaustion of the soil.

But farm-yard manure thoroughly restores to the

soil the power of producing the same succession of crops a second, a third, and a hundredth time: where it is applied in proper quantities it will fully cure the state of exhaustion, and often make a field more fertile than it ever was before.

The restoration of fertility by farm-yard manure cannot be attributed to the mixture of combustible materials (salts of ammonia and the substance of decaying saw-dust): for if these had a favourable effect, it must have been of a subordinate kind. The action of farm-yard manure most undoubtedly depends upon the incombustible ash-constituents of the plants which it contains.

In farm-yard manure the field actually receives a certain quantity of all the mineral ingredients which have been removed in the crops. The decline of fertility was in proportion to the removal of mineral constituents; the renewal of productiveness is in proportion to their restoration.

The incombustible elements of cultivated plants do not of themselves return to the soil, as the combustible elements return to the atmosphere from which they spring. The hand of man alone restores to the ground the conditions of the life of plants: in farm-yard manure wherein they are contained, the farmer, following a natural law, restores the lost power of production.

CHAPTER V.

THE SYSTEM OF FARM-YARD MANURING.

Questions to be solved—Experiments of Renning, their significance—Produce of unmanured fields—Influence of preceding crops, of the situation, and climatic conditions, on the produce—Each field possesses its own power of production—Large crops, their dependence and continuation—Closeness of the food of plants, what is meant thereby—The closeness of the particles of food in the soil is in proportion to the produce—Produce of corn and straw influenced by the relations of the assimilated food and by the conditions of growth; action of food supplied in manures—Potatoes, oats, and clover crops of the Saxon fields; conclusions drawn from them as to the condition of the fields—Produce of these fields from farm-yard manure; the increase of produce cannot be calculated from the amount of manure used—Restoration of the power of production of exhausted fields by the increase of the necessary elements of food present in the soil in minimum amount; advantageous use of farm-yard manure in this respect; explanation of the result- Action of manure as compared with quantity used: experiments—Rational system of cultivation—Depth to which the food of plants penetrates is dependent on the power of absorption of the soil; the Saxon fields considered in this respect; the power of absorption considered in manuring—Change produced in the composition of the soil by the system of farm-yard manuring; the different stages of this system, the final result—Examples of these stages in the Saxon experimental fields—Cause of the growth of weeds; remedies—The history of husbandry, what is taught by it—Present condition of European husbandry—Present production of the land compared with the earlier, conclusions—Continuation of production regulated by a natural law—Law of restoration; defective practice of it—Agriculture in the time of Charlemagne—Agriculture in the Palatinate—Corn fields in the valleys of the Nile and Ganges; nature provides in them for the restoration of food of plants—Practical agriculture and thelaw of restoration—The statistical returns of average crops afford an explanation of the condition of corn fields.

THE general observations in the preceding chapters on the mutual relations between the soil and plants, as also on the sources and nature of farm-yard manure, will, I hope, enable the reader to enter upon a thorough investigation of all those phenomena which are presented by the practice of farm-yard manuring. We have to consider how farm-yard manure increases the produce of a field; on which constituents of the manure its action depends; what quantity of farm-yard manure can be obtained from a field; and to what condition, after a series of years, a field can be restored by farm-yard manuring.

It will be understood that from this investigation we exclude all those effects of farm-yard manure which cannot be determined by measure and number; such, for instance, as its influence upon the looseness or cohesion of the soil, and its *heating* action, by means of the warmth resulting from the decay of its constituents in the ground.

The facts, to which this investigation extends, are derived from practical experience; and my selection of them has been materially facilitated by the comprehensive series of experiments made in the year 1851, at the instance of DR. RENNING, Secretary-General of the Agricultural Society in the kingdom of Saxony, by a number of Saxon agriculturists, with a view of 'ascertaining the action of so-called artificial manures under every variety of condition, for the purpose of more generally extending their application.' These experiments were continued to the year 1854, every series embracing a rotation of rye, potatoes, oats, and clover. The farmers were requested to try bone-dust, rape-cake, meal, guano, and farm-yard manure, each on a Saxon acre (= 1·36 English acre) of ground compared with an unmanured plot of the same size, and to determine the respective crops by weight.

Of all experiments of a similar nature which have been made in the course of several centuries, those which are expressly stated to have been undertaken 'without a direct scientific object' are of the highest scientific importance, not only for their very comprehensive character, but because they have resulted in fully establishing a number of facts which will for all time to come retain their validity as safe bases for scientific conclusions. Science owes the deepest gratitude to the excellent propounder of these inquiries, and to the worthy men who so zealously performed their task; the only thing to be regretted is, that the experiments upon unmanured plots were not carried out in all cases.

It is evident that the action of farm-yard manure upon a field can be properly estimated only if it is known beforehand what amount of produce the field will give without any manure: and first of all we shall consider the crops produced on five fields in five different parts of Saxony, in the four-year rotation above mentioned.

Crop.	Unmanured.				
	Cunnersdorf.	Mäusegast mixture.	Kötitz white clover.	Oberbobritzsch red clover.	Oberschöna grass.
	lbs.	lbs.	lbs.	lbs.	lbs.
1851. Rye					
Grain....	1176	2238	1264	1453	708
Straw ...	2951	4582	3013	3015	1524
1852. Potatoes..	16667	16896	18577	9751	11095
1853. Oats					
Grain....	2019	1289	1339	1528	1082
Straw ...	2563	1840	1357	1812	1714
1854. Clover-hay	9144	5583	1095	911	—

These results lead to the following considerations.

The term *unmanured*, as applied to these fields, is meant to designate the condition in which they were left at the end of a rotation by a succession of crops.

These fields had been manured at the beginning of the rotation; and had they been manured afresh, they would have produced the same crops as before. In the crops yielded by them in the manured state, the constituents of the soil and those of the manure had a certain definite share; if the fields had not been manured, the crops would have been smaller. Now if we attribute the increased produce during the course of the rotation to the supply of farm-yard manure, and suppose that the constituents of the farm-yard manure have been again removed in the crops, which is not true in all cases, then the field, at the end of the rotation, is in the same state in which it was at the commencement, before it had been manured. Accordingly, we may assume, without great risk of error, that the produce of different crops, which a plot of ground will yield in a new rotation without manuring, will be in proportion to the store of nutritive substances, ready for assimilation, which it contains in its natural state. Hence from the unequal products yielded by the two fields in that state, we may, with an approximation to truth, infer certain inequalities in the amount of food or in the condition of the fields.

Of course, inferences of this kind are admissible only within very narrow limits; for when we compare two fields which lie in the same or in different districts, we must remember that in each case various factors operate upon the products, making these unequal, even though the nature of the soil be otherwise identical.

If, for instance, two fields, both unmanured, are planted with one and the same cereal, it is by no means a matter of indifference, as regards the produce of corn and straw, what crop has preceded the cereal. If the last crop in the preceding rotation was clover on the one, oats on the other field, the results will vary, even though the condition of the soil in both was originally identical; and the produce reaped, in that case, indicates merely the state into which the field has been brought by the preceding crop.

In hilly districts, a northern or southern aspect makes a difference in the comparative character of two fields; so too does the height above the sea, on which the quantity of the fall of rain depends. A fall of rain received at a more favourable time by one field than by another makes a difference in the amount of produce, even though the condition of the soil be the same in both fields.

Lastly, in judging, in the manner indicated, of the state and condition of a field, the weather during the preceding year must be taken into account.

The crop produced by a field in a year is always the maximum crop which it can yield under the conditions given: under more favourable external circumstances, that is, with better weather, the field would have furnished a greater crop; under more unfavourable circumstances, a smaller, always corresponding to the condition of the soil.

By the production of larger crops, in consequence of favourable weather, the field loses a comparatively greater amount of nutritive substances, and the subsequent harvests show a decline; just as, on the other hand, deficient crops will act upon the yield of subsequent years, as a fallow year with half-manuring does, that is, the crops coming after bad years will turn out better, even in ordinary weather.

The relative proportions of corn and straw, in a crop of cereals, are altered by a continuance of dry or wet weather. Permanent wet, combined with a high temperature, favours the development of leaves, stalks and roots; and as the plant goes on growing, the materials intended for the production of seed are used for the formation of new shoots, and thus the seed crop is diminished.

Continuous drought, before or during sprouting time, produces the opposite effect; the store of formative matter accumulated in the roots is used in far greater proportion for the production of seed, and the relation of straw to corn is smaller than it would be in ordinary weather.

When all these circumstances are taken into account, the consideration of the produce obtained from unmanured fields in the Saxon experiments will leave only a few general points for further investigation.

The tabular statement of the result shows that each field has a power of production peculiar to itself, and that no two of them have produced the same amount of rye corn and straw, or potatoes, or oats and straw, or clover.

If we compare the numberless manuring experiments of the last few years, in which the crops obtained from unmanured plots were likewise taken into account, we see that this is a general rule admitting of no exception: no two fields have exactly the same productive power; nay, there are not even two plots in the same field which are identical in this respect. We need only look at a turnip field to see at once that every turnip differs in size and weight from the one growing next to it. This fact is so universally known and admitted, that in all countries where the land is taxed, the amount of the impost is assessed according to the quality of the soil, in some countries in eight classes, in others in twelve or sixteen.

Since, then, no two fields are alike in productive power, and every field must necessarily contain the conditions required for the production of the crops which it yields, it is clear that the conditions for the production of corn and straw, or of turnips and potatoes, or of clover or any other plant, are in no two fields alike: in one field the conditions for the production of straw preponderate over those for the production of grain, another is better suited for the growth of clover, and so on.

These conditions, according to their very nature, differ in quantity and quality. By conditions which can be weighed and measured, we of course mean no other than nutritive substances.

The crops reaped from a field afford no indication of the *quantity* of nutritive substances in the ground. Conseqfiently, the fact that the field at Mäusegast gave twice as much corn and one-third more straw than the one at Cunnersdorf, cannot lead to the inference that the former was upon the whole richer in these proportions in the conditions for the production of corn and straw; for we see that the Cunnersdorf field gave two years after, without manuring, one-half more oat-corn and straw than the field at Mäusegast, and in the fourth year above 60 per cent. more clover. Now some of the most important food elements of corn are as essential to clover as to the cereals; and the food elements of oats are identical with those of rye.

A larger crop of any of the cultivated plants given by one field over another merely indicates that the roots in the one field in their way downwards, have found and absorbed in certain portions of the soil more particles of the whole store of nutritive substances contained in it in an available state than the roots in the other field; but not that the total sum was greater in the one than in the other: for the field apparently poorer might in reality have contained a much larger total amount of nutritive substances than the other, only not in a condition available to the roots.

High returns are a sure sign that the nutritive substances of the soil are in a condition available to the roots; the *permanence* of high returns, and that alone, affords a safe criterion of the total store or *quantity* of nutritive substances in the ground.

The high returns yielded by one field above another result from this, that the particles of the mineral constituents lie nearer together in the one field than in the other: they depend upon the *closeness* of the nutritive substances. The following table may make this point clearer:—

Cunnersdorf, Mäusegast, Kötitz, Oberbobritzsch, Oberschöna.

Fig. I. 1851. WINTER-RYE.

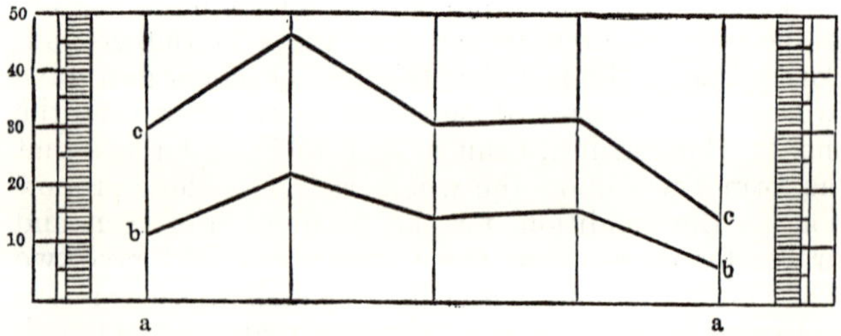

Fig. II. 1852. Potatoes.

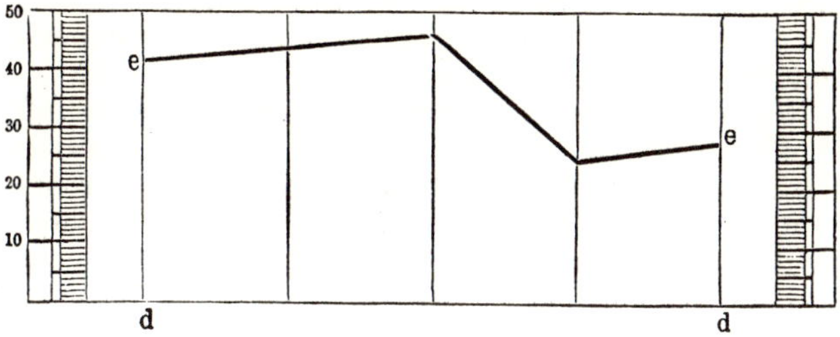

Fig. III. 1853. Oats.

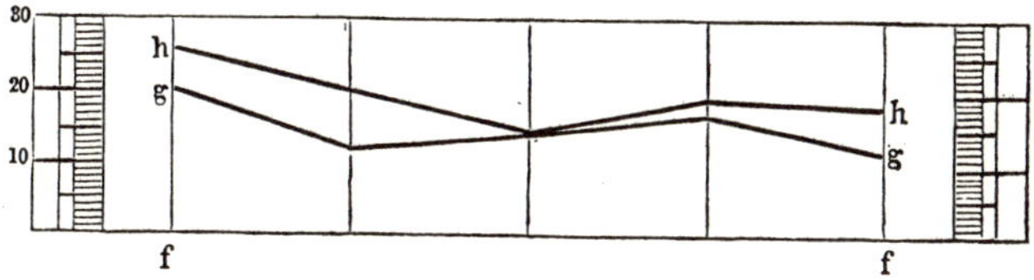

Fig. IV. 1854. Clover.

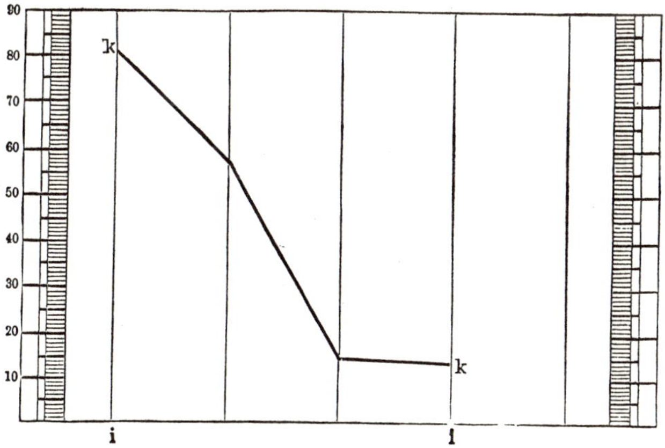

In Fig. I., the perpendicular lines *a b* represent the produce of grain, *a c* that of straw; in Fig. II., the lines *d e* the produce of potatoes; in Fig. III., the lines *f g* the produce of oat-corn, the lines *f h* that of oat-straw; in Fig. IV., the lines *i k* the produce of clover, on the unmanured plots of ground on which the experiments were made in Saxony.

Now if we assume that the roots of the rye and of the other plants, on the several fields, were of the same length and condition, it is quite certain that the roots of the cereals on the field at Mäusegast found, in their way downwards, much more nutriment than those in the Cunnersdorf field: the corn line is twice as high, and the straw-line one-third higher, in the former than in the latter.

With an equal number of plants, and an equal length of root, certain nutritive substances required by corn were twice as close in the Mäusegast as in the Cunnersdorf field. The line in Fig. IV. representing the produce of clover is ten times as high for Cunnersdorf as for Oberbobritzsch, which means that the nutritive substances required by clover were ten times as far asunder in Oberbobritzsch as in Cunnersdorf.

In comparing the produce of several fields, the *closeness* of the nutritive substances in the soil is in inverse proportion to the height of the lines in the table indicating the amount of produce.

The longer the lines, the closer are the nutritive substances in the various soils; the shorter the lines, the more widely asunder do the substances lie.

For instance, the lines indicating the produce of potatoes at Kötitz and Oberbobritzsch are as 18 to 9; the potato crop at Kötitz was twice as high as that at Oberbobritzsch. Hence it follows that the distance between the nutritive substances was in inverse ratio, that is, as 9 to 18; in the field at Kötitz they were twice as close together as in the other.

This mode of viewing the matter is calculated to lead, in many cases, to more definite ideas respecting the cause of the exhaustion of a field.

The corn and potato crops, for instance, took away phosphoric acid and nitrogen from the arable surface soil at Mäusegast, and the barley plant next in rotation, which likewise draws its nutriment from the surface soil, found in the third year much less nutriment than the rye plant which had preceded it.

The elevations of the lines *a b* (Fig. I.) and *f g* (Fig. III.), taken inversely, show how much relatively greater has become the distance between the particles of the nutritive substances for the barley plant. The barley-corn requires for its formation the same nutritive substances as the rye-corn. Now, as the produce of the rye-corn was to that of the barley-corn in the proportion of 22 : 12, this means, taken inversely, that the distance between the nutritive substances for the barley-corn had increased from 12 to 22.

In the third year, the roots of the barley, for the same length, found scarcely half as much nutriment for grain as the rye had found.

This exposition is not intended to supply a standard for measuring the distances between the available particles of nutritive substances in the ground, but merely to define more accurately what is meant by the exhaustion of land. The farmer who has a clear view of the causes upon which depends the reduction of crops by continuous cultivation, will thereby the more easily find out and apply the means to make his field as productive as before, and, if possible, even to increase its fertility.

Beside the general differences of all the crops in the Saxon experiments, we are further struck with the inequality in the proportion of corn and straw.

To 10 parts by weight of corn, the yield of straw was respectively—at Cunnersdorf 25 parts by weight, at Kötitz 23, at Oberschöna only 21, and at Mäusegast only 20.

A more careful examination of the table shows that the difference is mainly in the produce of corn.

The fields at

	Cunnersdorf.	Kötitz.	Oberbobritzsch.
Yielded in straw	2951 lbs.	3013 lbs.	3015 lbs.

that is, within a few pounds, the same quantity of *straw*, while the amount of *corn* was in

Cunnersdorf.		Kötitz.		Oberbobritzsch.
11	:	12	:	14

In investigating the reasons for this inequality in the produce of corn, we discover at the same time the causes of the difference in the proportion between the corn and straw.

It is necessary to remember that what is called *straw* (i. e. the leaves, stalks, and roots) is formed from the albumen of the cereal seeds, that is, from the constituent elements of the seeds; and, further, that these parts of the plant are the organs for the reproduction of these same seed constituents.

The production of the straw always precedes the formation of the grain; and that portion of the seed elements which serves to form the organs of the plant cannot be used to make seed: or, the more seed-constituents are turned into straw-constituents within the appointed time of growth, the fewer will remain at the close of that period for the formation of seed (see p. 63).

Before the period of flowering, all the seed-constituents go to form straw; after that period, a division takes place.

Therefore, if all other conditions of soil and weather are equally favourable, the quantity of straw will depend upon the amount of seed-constituents needed for the formation of straw.

The quantity of corn depends upon the residue of seed-constituents in the whole plant, which are no longer required for the multiplication and enlargement of leaves, stalks, and roots.

Let K represent that portion of the corn constituents that may be formed into seed; aK the other fraction of the same substances, which remain as constituents in the straw; and St the other constituents comprised in the straw: so that

K=(phosphoric acid, nitrogen, potash, lime, magnesia, iron),
aK=a fraction of K,
St=(silicic acid, potash, lime, magnesia, iron);

then the nutritive substances which the plant has absorbed from the soil, may be thus expressed:—

$$(K + aK\ St).$$

This expression, therefore, means that the roots of the cereal plant must have absorbed from the earthy particles in contact with them a certain proportion of nutritive substances for the production of leaves, roots, and stalks, and after this an additional amount of several of the same constituents for the formation of grain. The total produce is, of course, dependent upon the sum of the K and S*t* constituents, which the soil is able to supply to the plants during the natural period of growth.

The ratio between corn and straw results from a division of the K and S*t* constituents in the plant itself, and depends upon the relative proportion of the K and S*t* constituents in the soil, as also upon the action of external causes favouring the production of corn or straw.

When the quantity of K constituents in the ground decreases, less grain will be produced; but it is only in certain cases that this will exercise any influence upon the produce of straw.

When the quantity of S*t* constituents in a field is increased, the enhanced conditions for the formation of leaves, stalks, and roots, must injure the crop of grain, if the amount of *a*K required for the additional formation of straw is taken from the store of K contained in the soil.

If one of two fields is poorer in K but richer in S*t* constituents than the other, the former may give the same, perhaps even a larger, amount of straw, than the latter, but its produce of corn will necessarily be less.

A similar increase of straw, at the expense of grain, takes place when the state of the weather is more favourable for the formation of leaves, stalks, and roots, than for grain. The period of growth is thus prolonged, and the plant then takes up more of the S*t* constituents, which are usually in excess; for the assimilation of these, a certain additional quantity of the K constituents is consumed, which would otherwise have served to form seed.

Let *st* represent the additional supply of S*t* constituents afforded by the soil under these circumstances, and *ak* the additional portion of K converted into straw-constituents; then the alteration in the produce may be expressed as follows:—

$$\overset{\text{Corn.}}{(\mathrm{K} - ak)} + \overset{\text{Straw.}}{(a\mathrm{K}\ \mathrm{S}t + ak\ st)};$$

that is, the produce of straw increases, while that of grain diminishes. It is also evident, that where the S*t* constituents are in excess and the amount of K constituents is increased, then if K is proportionately deficient there will be an increase in the produce of straw, and if K is proportionately increased there will be a larger produce both of corn and straw.

As the constituents of K, with the exception of nitrogen and phosphoric acid, are also constituents of S*t*, this accession of produce in the field under consideration will be also effected either by a supply of phosphoric acid, or of nitrogen, or both together.

If by this supply the closeness of the K particles in the ground, or of the phosphoric acid and ammonia particles, is doubled, then under the most favourable circumstances the harvest may be doubled by the supply of K.

If, on the other hand, the soil is deficient in S*t* constituents, any increase of nitrogen or phosphoric acid in the ground will fail to exercise the slightest influence upon the crop.

It results from this, as a matter of course, that the absolute or relative amount of straw, given by a field in a crop of corn, will furnish no proof of the S*t* constituents in the soil: since, though two fields may be equally rich in these constituents, the produce of straw depends upon the quantity of K constituents in the ground: hence the field which is richer in K, will, under like circumstances, give a larger crop of straw.

The fact, therefore, that the fields at Cunnersdorf and Oberbobritzsch yielded a like amount of straw, cannot lead to the inference that these fields contained an equal quantity of S*t* constituents, since the corn crops show that the quantities of K were unequal. The harvests exhibited the following proportions:—

In Cunnersdorf as	(11) K : (29) *a*K S*t*.
" Kötitz as	(12) K : (30) *a*K S*t*.
" Oberbobritzsch as	(14) K : (30) *a*K S*t*.

As before remarked, the constituents represented by the symbols K and S*t* differ merely in this, that K comprises nitrogen and phosphoric acid, while the other constituents of K are common to both; hence the difference in the corn crops of these three fields results mainly from the fact, that the roots of the corn found in the soil at Kötitz $\frac{1}{11}$ and at Oberbobritzsch $\frac{3}{11}$ more phosphoric acid and nitrogen in an available condition than at Cunnersdorf.

If the question is asked, how much phosphoric acid and nitrogen must be added to the field at Cunnersdorf in order to make the crop of corn equal to that of Oberbobritzsch, it would be a mistake to suppose that an increase of $\frac{3}{11}$ would be sufficient; for the augmentation of the produce of corn is materially influenced by the S*t* constituents, the quantity of which varies greatly in different soils and has not been ascertained.

By the addition of nitrogen and phosphoric acid, a certain quantity of the accumulated S*t* constituents are rendered effective or available, which before were not so; but while the produce of straw increases, not $\frac{3}{11}$, but less of nitrogen and phosphoric acid remain over for the formation of seed; the exact quantity is limited by the total amount of transformed S*t* constituents.

The closeness of the S*t* constituents in different soils may, however, be approximately ascertained from the relative proportion of corn and straw obtained from a plot manured with phosphoric acid and nitrogen, and from an unmanured plot respectively.

If the unmanured plot yields corn and straw in the proportion of 1 : 2·5, and the manured plot gives a larger crop in which the corn is to the straw as 1 : 4 (straw being in greater proportion), it is evident that the S*t* constituents preponderate in the latter field; and a much larger quantity of phosphoric acid and nitrogen would have to be supplied in order that the field, correspondently with its amount of S*t* constituents, might produce the same relative proportion of corn and straw as, for example, the land at Oberbobritzsch.

It is a very essential part of the farmer's business to study the nature of his field, and to discover which of the nutritive substances, useful to plants, his land contains in preponderating quantity: for thus he will know how to make a right selection of such plants as require for their developement a superabundance of these constituents; and he will obtain the greatest profit from his field, when he knows what nutritive substances he must supply in due proportion to those which are already in abundance.

Two fields, in which the total amount of nutriment is unequal, but the relative distribution of the substances is the same, will produce crops differing in quantity, but agreeing in the relative proportion between corn and straw.

Such a relation, for example, exists between the field at Oberbobritzsch and the field at Mäusegast. If the crop of corn and straw in the former is expressed by $K + aK\ St$, the crop in the latter $= 1\frac{1}{3}K + 1\frac{1}{3}aK\ St$.

The fields are evidently cultivated in both places with great care and skill, and the soil is so uniformly mixed, that when we know the corn and straw crop of the one, and the straw crop of the other, we can calculate the corn crop of the latter from the above formula.

Potatoes, 1852.—In the subjoined table, the vertical lines show the potato crops from five different fields in the year 1852.

1852. POTATOES.

Cunnersdorf, Mäusegast, Kötitz, Oberbobritzsch, Oberschöna.

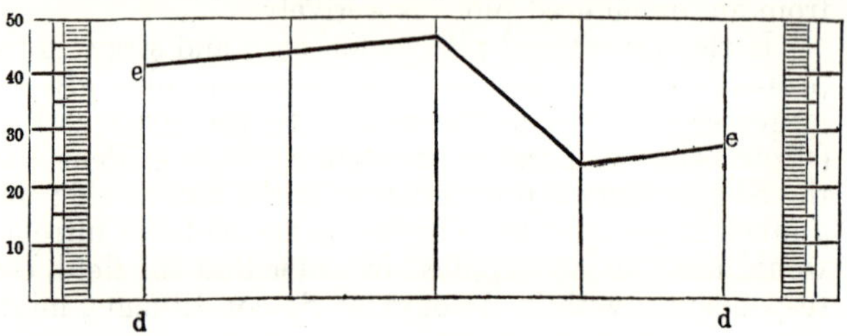

The potato plant draws its principal constituents from the arable surface soil, and from a somewhat deeper layer than the rye plant; and the crops reaped show the condition of the layers more accurately than could be ascertained by chemical analysis.

In the fields at Mäusegast and Cunnersdorf the nutritive substances available for the potato plant were about equally close; in Kötitz they were one-ninth closer to each other; at Oberbobritzsch they were twice as far asunder; while at Oberschöna they were one-fifth closer than in Oberbobritzsch.

The largest potato crop was obtained from the field at Kötitz. Potash (for the tubers) and lime (for the herbaceous parts) are the predominant constituents of the potato plant: but a certain amount of nitrogen and phosphoric acid is as necessary for the developement of the potato as it is for cereals; and the effective quantity of the transmuted potash and lime is essentially determined by the phosphoric acid and nitrogen absorbed at the same time. Where one of the two latter elements which, as we have remarked, are equally constituents of cereals, is deficient in the soil, the potato crop will be proportionate to the available quantity of these two substances, and the greatest excess of potash or lime in the soil will have no influence whatever upon the amount of the produce.

The arable surface soil of the field at Oberbobritzsch is much richer in phosphoric acid and nitrogen than that of the Kötitz field; yet the potato crop yielded by the former was only half that given by the latter.

Accordingly, nothing can be more certain than that the field at Oberbobritzsch contained much less potash or lime in an available state, than the Kötitz field; and by manuring with lime alone, or with wood-ashes (potash and lime), it might readily be ascertained in which of the two substances the ground was deficient.

But from the inferior potato crop given by the field at Cunnersdorf, we cannot infer that it was poorer in potash or lime than the field at Kötitz; the latter decidedly contained, as the preceding corn crop shows, somewhat more phosphoric acid and nitrogen than the field at Cunnersdorf: consequently, the larger potato crop at Kötitz may have been mainly owing to the greater quantity of these two elements contained in it. Even if the field at Cunnersdorf had been still richer in potash and lime than the Kötitz field, yet after all, under the given conditions, it would have produced a smaller crop of potatoes.

Oats, 1853.—The oat plant derives part of its nutriment from the arable surface soil, but sends its roots, when the soil permits, much deeper than the potato; it possesses, so to speak, a higher power of vegetation than the rye plant, and in the faculty of appropriating nutriment resembles weeds.

1853. OATS.

Cunnersdorf, Mäusegast, Kötitz, Oberbobritzsch, Oberschöna.

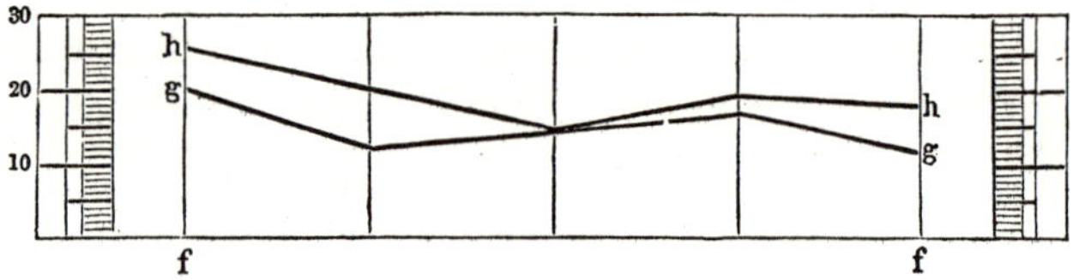

The point which most strikes us in this table is the great inequality in the produce of two cereal plants grown successively on the same unmanured soil.

The field at Cunnersdorf, which next to that at Oberschöna had given the lowest crop of rye-corn and straw, yielded in the third year the largest produce of oat-corn and straw.

The difference in the condition and closeness of the nutritive substances in the lower layers of these fields is undeniable. The field at Cunnersdorf was poorer in the upper layers, but went on increasing downwards in the amount of substances nutritive to the corn plant; the other fields decreased downwards.

The returns of the field at Mäusegast for the year 1853 refer to barley and not to oats: hence they afford no conclusion as to the condition of the deeper layers, from which the oat plant derives its food: but they show the state into which the arable surface soil had been brought by the preceding corn crop. Owing to the abstraction of phosphoric acid, and perhaps of nitrogen, the yield of barley-corn was much less than might have been expected from the soil, judging by the preceding rye crop; and a small supply of superphosphate or guano would have greatly increased the produce of barley on this field.

Clover, 1854.—The clover crops in the fourth year afford an insight into the condition of the deepest layers from which plants draw their food.

1854. Clover.

Cunnersdorf, Mäusegast, Kötitz, Oberbobritzsch, Oberschöna.

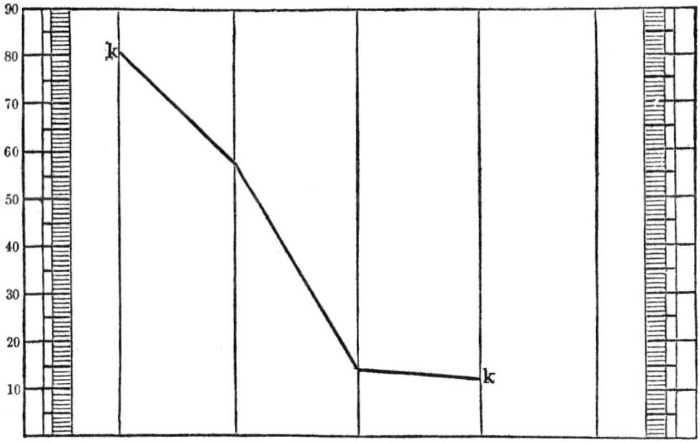

The produce of clover at Cunnersdorf was nearly twice as large as at Mäusegast, and ten times greater than at Oberbobritzsch; and it is beyond doubt, that these unequal crops must have corresponded to unequal amounts in the soil of substances nutritive to the clover plant.

The substances required by the clover plant, in respect of quantity and relative proportion, are very nearly the same as for the potato plant (leaves, stalks, and tubers included): and if clover still yields good crops upon a soil wherein potatoes thrive but imperfectly, this is chiefly owing to the wider root-ramification of the clover plant. There are scarcely any two other plants which so clearly indicate the layers of the soil assigned to them by nature, for the absorption of their nutriment.

If potatoes are planted in trenches two feet deep, and if these are filled up in proportion as the plant grows, so that at last the earth in the trench is on the same level with the arable surface, it is always found that the tubers are formed only in the topmost layer, none at a greater depth, and not more in number than if the seed-potatoes had been planted only 1½ or 2 inches deep in the arable surface soil: and on gathering the crop it is observed that the roots below the arable surface have died away.

With clover, the case is reversed; and although the arable surface soil at Kötitz, for example, is decidedly richer in substances nutritive for clover than that in Cunnersdorf (yielding a potato crop higher by one-eighth), this had no effect upon the clover, which receives its principal nutriment from the deepest layers of the soil.

We now proceed to an analysis of the returns which were obtained, in the Saxon experiments, by employing farm-yard manure upon the plots of the same fields, the crops of which in their unmanured state we have just been considering.

Produce, per Saxon acre, of the fields dressed with farm-yard manure.

	Cunnersdorf.	Mäusegast.	Kötitz.	Oberbobritzsch	Oberschöna.
	cwt.	cwt.	cwt.	cwt.	cwt.
Farm-yard manure...	180	194	229	314	897
1851.	lbs.	lbs.	lbs.	lbs.	lbs.
Rye corn	1513	2583	1616	1905	1875
" straw....	4696	5318	4019	3928	3818
1852.					
Potatoes.....	17946	20258	20678	11936	16727
1853.					
Oat corn.....	2278	1649	1880	1685	1253
" straw....	2992	2475	1742	1909	2576
1854.					
Clover-hay....	9509	7198	1232	2735	0*

Increase by farm-yard manure over unmanured plots. (See p. 186.)

	Cunnersdorf.	Mausegast.	Kotitz.	Oberbobritzsch.	Oberschöna.
	lbs.	lbs.	lbs.	lbs.	lbs.
1851.					
Rye corn	337	345	352	452	1167
" straw....	1745	736	1006	915	229
1852.					
Potatoes.....	1279	3362	2101	2185	5632
1853.					
Oat corn.....	369	360	541	157	171
" straw....	429	635	385	97	862
1854.					
Clover-hay ...	365	1615	137	1824	—*

Here, again, what strikes us first is that the returns from all the fields were different from one another, and that apparently they did not bear the most remote relation to the quantity of manure applied.

Nothing can be more certain than the fact that a field, exhausted by cultivation, will yield larger returns if dressed with farm-yard manure than if unmanured:

* The clover crop failed from excessive wet.

now, taking the increase to be caused by manure, it is natural to suppose that the same quantity of manure would produce the same increase upon different fields. The following table, however, shows that the same quantity of manure, upon the Saxon fields, produced results which differed very considerably.

One hundred cwt. of farm-yard manure gave increased produce.

	Cunnersdorf.	Mausegast.	Kötitz.	Oberbobritzsch.	Oberschöna.
	lbs.	lbs.	lbs.	lbs.	lbs.
1851–53. Winter rye & oats	1539	1070	988	515	501
1852. Potatoes	720	1723	917	696	628
1854. Clover	203	832	60	628	—

No one looking at these numbers could divine that they were intended to represent the effects produced upon five different fields by an equal quantity of the same manure, and that too the universal manuring agent.

Neither in the crop of rye-corn and straw, nor in that of potatoes, oats, and clover, is there the slightest resemblance or correspondence; still less is it possible to discover what amount of manure has been instrumental in producing the increased crops.

The same quantity of farm-yard manure gave, in the years 1851 and 1853, at Mäusegast double, at Cunnersdorf three times, the increase of cereal crops, corn and straw together, that was obtained at Oberbobritzsch: the increase of the potato crop at Mäusegast was twice as large as in Kötitz; of clover, four times more in Mäusegast than in Cunnersdorf; and in Oberbobritzsch, ten times as much as in Kötitz.

The enormous quantity of farm-yard manure put upon the field at Oberschöna failed to produce anything like the crop obtained from the unmanured field at Mäusegast.

The composition of farm-yard manure, as we know from numerous analyses, is on the whole so much alike in all places, that we may suppose without great risk of error that in 100 cwt. of farm-yard manure every field receives the same nutritive substances and in the same quantities.

The constituents of farm-yard manure act everywhere in the same way upon the soil or the earthy particles. Now this apparently involves an irreconcilable contradiction with the fact that the increase obtained by it is nevertheless everywhere different, and that the dung-constituents supplied will, on one field, set in motion and render available to the cereal or potato plants growing on it, twice or three times as many elements of food as on another field.

This fact does not refer to the Saxon fields alone, but applies generally. Nowhere, in no country, do the crops obtained by farm-yard manuring on different fields ever correspond, as the following table of the average produce of divers crops in different provinces of the kingdom of Bavaria will show.

AVERAGE CROPS IN BAVARIA.

(*Seuffert's Statistics.*)

One day's work yields average produce in bushels.*

	Wheat.	Rye.	Spelt.	Barley.	Oats.
Upper Bavaria	1·70	1·80	3·40	1·90	2·31
Lower Bavaria	2·50	1·80	3·40	1·90	2·31
Upper Palatinate and Ratisbon	1·45	1·40	2·70	1·75	1·85
Upper Franconia	1·20	1·30	2·20	1·50	1·75
Middle Franconia	1·65	1·40	3·50	1·65	2·25
Lower Franconia and Ashaffenburg	1·70	1·75	2·50	2·00	2·75
Suabia and Neuburg	1·80	2·00	5·0	2·30	3·50
Palatinate	2·70	2·60	4·80	3·75	3·90

* 1 *Hectolitre*	*weighs on an average* Zollverein weight.	1 *Bavarian bushel.* Zollverein weight.
Wheat	146 lbs.	330—345 lbs.
Barley	128 "	290—300 "
Rye	140 "	318—325 "
Oats	88 "	200—300 "
Spelt (in the husk)	79 "	174—220 "

According to this scale, the weight of a Prussian bushel of wheat is

The crops produced by farm-yard manuring differ not only in every country, but even in every locality; and, strictly speaking, every field dressed with farm-yard manure yields an average produce of its own.

The action of farm-yard manure upon the increase of produce is intimately connected with the condition and composition of the soil; it varies, therefore, in different fields, simply because the composition of the soil varies.

To understand the action of farm-yard manure, it is necessary to remember that the exhaustion of a field arises from the loss of a certain amount of nutritive constituents, at the end of a rotation, inflicted upon the soil by preceding crops, which of course leave less available food in the soil for the following crops.

However, the loss of each individual constituent has not the same effect upon the exhaustion of the soil.

The loss of lime which a calcareous soil suffers by a cereal or by clover, matters little to the growth of a succeeding plant that requires large quantities of lime to thrive well. The same applies equally to the loss of potash, magnesia, iron, phosphoric acid, nitrogen, on fields severally abounding in potash, magnesia, iron, phosphoric acid, or ammonia. Where a soil is abundantly provided with one of the mineral constituents, the amount of that constituent removed by the crops is so small a fraction of the whole mass, that the effect of the diminished store is not appreciable from one rotation to another.

But practical experience shows that the crops do decrease from one rotation to another, and that the land requires a fresh supply of certain ingredients by manuring, if it is again to produce as large harvests as before.

Now, as a supply of lime cannot be expected to restore the fertility of an exhausted field where lime constitutes the principal bulk of the soil, just as little as a supply of potash or phosphoric acid to a field abounding in potash or phosphoric acid, it is easy to understand

83 lbs., and that of the English quarter 425 lbs., 100 lbs. (Zollv. weight =110·2 lbs. avoir.).

that where the productive power of an exhausted field is restored, the fertilising effect is to be attributed simply to the manure returning to the field those elements of food which the soil originally contained in the least proportion, and of which it has accordingly lost, by the preceding crops, comparatively the largest fraction.

Every field contains a *maximum* of one or several, and a *minimum* of one or several, other nutritive substances.. It is by the *minimum* that the crops are governed, be it lime, potash, nitrogen, phosphoric acid, magnesia, or any other mineral constituent; it regulates and determines the amount or continuance of the crops.

Where lime or magnesia, for instance, is the minimum constituent, the produce of corn and straw, turnips, potatoes, or clover, will not be increased by a supply of even a hundred times the actual store of potash, phosphoric acid, silicic acid, &c., in the ground. But a simple dressing with lime will increase the crops on a field of the kind, and a much larger produce of cereals, turnips, and clover will be obtained by the use of this agent (just as is the case by the application of wood-ashes on a field deficient in potash) than by the most liberal use of farm-yard manure.

This sufficiently explains the dissimilar action upon different fields of so composite a manure as farm-yard dung.

Only those ingredients of farm-yard manure which serve to supply an existing deficiency of one or two of the mineral constituents in the soil act favourably in restoring the original fertility to a field exhausted by cultivation; all the other ingredients of the manure, which the field contains in abundance, are completely without effect.

A field rich in straw-constituents cannot be made more productive by manuring with straw-constituents in the dung, whereas these constituents will prove most efficacious on fields deficient in them.

If two fields have the same abundance of straw-constituents, but are not equally rich in corn constituents, the same supply of farm-yard manure will not produce, by any means, equal crops of corn, because these must bear a relation to the corn-constituents supplied in the manure. Of these, both fields received the same amount in the same quantity of manure; but as the one field, of itself, was richer in corn-constituents than the other, the poorer of the two must receive much more manure to make it produce as large crops as the other.

A comparatively small quantity of superphosphate will, on a field of the kind, serve to increase the produce to a much greater extent, than the most liberal use of farm-yard manure.

Upon a field deficient in potash farm-yard manure acts by the potash contained in it; upon a soil poor in magnesia or lime, by its magnesia or lime; upon one poor in silicic acid, by the straw in it; upon land poor in chloride or iron, by the chloride of sodium, chloride of potassium, or iron contained therein.

This fact accounts for the high favour in which farm-yard manure is held by practical farmers. As the dung of the farm-yard contains, under all circumstances, a certain quantity of each of the mineral constituents withdrawn from the soil by the crops grown on it, its action is universally beneficial. It never fails to produce the desired effect, and thus spares the practical man the trouble of devising more suitable and equally efficacious means for keeping up the fertility of his fields, with a less profuse expenditure of money and labour, or of raising his land, without additional outlay, to the highest attainable degree of fertility compatible with its composition.

It is well-known in practice, that the produce of many fields may be increased by manuring with guano, bone-dust, rape-cake, and other substances containing only certain constituents of farm-yard manure; and their operation is explained, in effect, by the doctrine of *minimum*, which I have just laid down.

But as the practical farmer is not acquainted with the law which regulates the operation of these manuring agents as affecting the increase of produce, he can, of course, have no correct notion of their rational, which means their truly economical, use; he puts on his land too much, or too little, or chooses the wrong agent. The error of employing too little manure needs no explanation; for every one knows that the right proportion of manure will, with exactly the same labour and at a trifling additional outlay, ensure the maximum produce of which the land is capable.

The error of using too much manure arises from the mistaken notion that the action of manures is proportionate to the quantities in which they are applied; this is true up to a certain limit, but beyond this all the manure applied is simply thrown away, as far as any fertilising action is concerned.

A manuring experiment made by Mr. J. Russell, of Craigie House ('Journal of the Royal Agr. Soc. of England,' vol. xxii. p. 86), may, perhaps, serve to illustrate our meaning. In this experiment a field was divided into a number of plots of three rows each, all planted with turnips, some of the plots being left unmanured, the remainder dressed severally with different manuring agents, among others with superphosphate (bone-ash dissolved in sulphuric acid). The produce, calculated per acre, was as follows:—

Produce per acre.

No. of plots.		Cwt.	
I.	Unmanured	340	turnips (Swedes).
II.	"	320	"
V.	Manured with 5 cwt. of superphosphate	535	"
VI.	" 5 " "	497	"
VII.	" 3 " "	480	"
VIII.	" 7 " "	499	"
IX.	" 10 " "	490	"

As shown by the difference of 20 cwt. in the produce of the unmanured plots, the condition of the soil and the store of mineral constituents differed, to some extent, in different parts of the field. Other experiments, which we cannot describe more particularly, showed that the soil was poorer in the centre of the field than on the sides.

The one great fact most clearly proved by the above table of produce is, that 3 cwt. of superphosphate gave nearly the same crop of turnips as 5 cwt.; and that a further increase of the manure to 10 cwt. produced no additional increase of the crop.

No steps were taken, in these experiments, to ascertain which of the constituents of superphosphate of lime had the principal share in increasing the produce of the field. Magnesia and lime, as well as sulphuric and phosphoric acid, are equally indispensable elements of food for the turnip plant; and I have observed that by manuring with gypsum and a little common salt or with phosphate of magnesia, a field will be made to give more abundant crops than by employing superphosphate of lime, although the latter unquestionably proves the most effective manure for most fields.

To apprehend these facts correctly, we must remember that the law of the *minimum* does not apply to one constituent alone, but to all. Where, in any given case, the crops of any plant are limited by a minimum of phosphoric acid in the field, these crops will increase by augmenting the quantity of phosphoric acid up to the point at which the additional phosphoric acid bears a proper proportion to the next minimum constituent in the soil.

If the additional phosphoric acid exceeds the corresponding quantity, for instance, of potash or ammonia in the soil, the excess will prove of no effect. Before the supply of phosphoric acid the available quantity of potash or ammonia was a little larger than the amount of phosphoric acid in the soil, and the excess of the alkalies was ineffective until the phosphoric acid was supplied; similarly the excess of phosphoric acid must remain just as inoperative, as previously the excess of potash.

Whilst the produce before was proportionate to the minimum of phosphoric acid, it is now in proportion to the minimum of potash or ammonia, or both alkalies. A few experiments made on Mr. Russell's field might have settled the question. Had potash or ammonia been the minimum, after manuring with superphosphate, a suitable supply of potash or ammonia, or both, would have increased the produce. In this same series of experiments, 6 cwt. of guano, corresponding to 2 cwt. of superphosphate, gave a crop of 630 cwt. of turnips, or 130 cwt. more than the superphosphate; but it is left in doubt whether this increase was attributable to the potash or the ammonia in the guano.

To return to our Saxon experiments. If we look at the different quantities of dung applied severally on the five fields, we are naturally led to inquire the reason of this diversity.

The most *feasible* answer, perhaps, is, that the farmer gives as much manure as he has at his disposal, or that he regulates the quantity according to certain facts. If he has found by experience that a certain quantity of farm-yard manure will restore his land to its original fertility, and that more copious manuring will fail to give larger crops, in proportion to the additional supply, or to the cost incurred in collecting the manure, he will stop at the smaller quantity.

Hence it cannot be regarded as a mere accident that the farmer at Cunnersdorf contented himself with 180 cwt. of farm-yard manure, while the farmer at Oberbobritzsch laid 314 cwt. upon his field.

But if the quantity of manure to be applied is not dependent upon chance or caprice, but is regulated by the object in view, it is manifest that the proceedings of the farmer are governed by a law of nature unknown to him, except by its effects.

It is in the composition and condition of the soil that we must seek the law which regulates the quantity of farm-yard manure required, at the outset of a fresh rotation, to restore a field to its former fertility; and it is not difficult to see that this quantity must always be proportionate to the effective dung-constituents already present in the soil; a field largely abounding in them takes less manure than a poor field to give the same increased produce.

Now, as farm-yard manure owes its most active

constituents to clover, turnips, and the grasses, the inference is pretty clear that the quantity of this manure required on a field is in an inverse ratio to the produce of clover, turnips, or grass, which the field can give when unmanured.

The Saxon experiments show that this inference cannot be far from the truth, in one respect at least; for on comparing the produce of clover given by the unmanured plots with the quantity of farm-yard manure applied, we find:—

Clover crops in 1854.

	Cunnersdorf.	Mäusegast.	Kötitz.	Oberbobritzsch.	Oberschöna.
Pounds..	9144	5583	1095	911	—

Quantity of manure applied in 1851.

Cwt.....	180	194	229	314	897

The field at Cunnersdorf which contained the largest store of dung-constituents received the smallest; the field at Oberbobritzsch which gave the smallest crop of clover, the largest quantity of farm-yard manure.

The crop of clover, however, is not the only factor to determine the amount of farm-yard dung required for manuring; for one of the clover-constituents, silicic acid, which is indispensable to the cereal plants, is present only in trifling proportion, and hence the quantity of farm-yard manure (straw-manure) must bear a definite ratio to the quantity of straw-constituents already present in the ground.

If, in the Saxon experiments, we compare the increased produce of corn and straw obtained from the fields manured with farm-yard dung, we find:—

Increase of produce by farm-yard manuring, per acre.

	Cunnersdorf.	Kötitz.	Oberbobritzsch.
Quantity of farm-yard manure....	180 cwt.	229 cwt.	314 cwt.
Corn........................	347 lbs.	352 lbs.	452 lbs.
Straw........................	1743 "	1006 "	914 "

The field in Cunnersdorf, manifestly the richest in substances nutritive for straw, gave the largest straw-crop, although it had received the smallest quantity of farm-yard manure. In the increased produce, corn was to straw as 1 : 5, clearly showing that sparing application of straw manure was the proper course to pursue here. This fact readily explains also why the field at Oberbobritzsch, comparatively poorer in straw-constituents, required 85 cwt. of farm-yard manure more than the Kötitz field, to enable it to maintain, in its increased produce, the same proportion of corn and straw (1 : 2) as in the crop from the unmanured plot.

These considerations might, perhaps, lead the practical farmer to the conviction that he is, after all, not much of a free agent in the cultivation of his fields, and that the 'facts and circumstances' which guide him in his proceedings are simply laws of nature, of whose existence he has scarcely any conception. In truth, it may be said that the agriculturist is a free agent only in his wrong-doings. If he acts in accordance with his own interest, he must allow himself to be guided, even though unconsciously, by the condition of his land; and the only matter for wonder is, how far the man of 'experience' has succeeded in this way.

A system of farming, to be called truly *rational*, must be exactly suited to the nature and condition of the soil; for it is only when the rotation of crops or the mode of manuring is conformable to the composition of the soil, that the farmer has a sure prospect of realising the highest possible returns from his labour or from the capital invested.

Now considering, for instance, the great difference in the condition of the soil at Cunnersdorf and Oberbobritzsch, it is self-evident that the same rotation of crops which suits the one field, will not answer equally well for the other.

If farmers would only make up their minds to acquire by experiments on a small scale,* an accurate knowledge of the productive power of their land for certain kinds or classes of plants, a few more experi-

* In a field of pretty uniform composition, experiments of this kind may be made with flower pots sunk in the earth.

ments would readily enable them to discover what nutritive substances their land contains in minimum proportion, and what manuring agents ought to be applied to ensure the production of a maximum crop.

In matters of this kind the farmer must pursue his own course, and the proper course is the one that will most fully secure the object he has in view; he must not put the least faith in the assertion of any foolish chemist, who wants to prove to him analytically that his field contains an inexhaustible store of this or that nutritive substance. For the fertility of a field is not proportionate to the quantity of one or several food elements analytically shown to exist in it, but to that fraction of the total nutritive substances which the field is able to give up to the plants; and the only means of determining that fraction is by the plant itself. The most that chemical analysis can do is to supply a few data for comparing the condition of two fields. The experiments made by the beet-root growers on the extensive tract of land in Russia, known as the *Tschernosem* or 'Black soil,' whose fertility for corn plants is proverbial, show that this earth, though analytically proved to contain upon the whole, to the depth of twenty inches, 700 to 1000 times the quantity of potash required for a full beet-root crop, is, after three or four years' cultivation, so exhausted, that without manuring it will no longer yield a remunerative crop of beetroot.*

In the produce of cereals there is only *one* proper proportion between grain and straw; but the unfavour-

* With regard to the general opinion about the abundance and inexhaustibility of potash in land, the following announcement, in the 'Badische Centralblatt für Staats und Gemeinde-Interessen,' May 1861, is not without interest. 'In the District of Bretten.—The contracts which usually take place in the early part of the year for the cultivation of beetroot, are now fully open for competition in this district, and for good articles 30 francs the cwt. are offered this year, whereas last year only 26 francs were paid. Notwithstanding this rise of prices, and the premiums offered for superior roots, not many transactions have been concluded. The reason of this is quite intelligible, for the very injurious effects resulting to land on which this product has been cultivated, are too well known.' The effects must have reference to fields which had been adequately manured, for otherwise no profitable returns can be expected.

able proportions are many. It is clear that the mass and extent of the organs for the formation of grain (in other words, the bulk of the straw) must bear a definite relation to the product, that is, to the quantity of grain produced: any excess or deficiency in the amount of straw must always act injuriously upon the grain crop.

When it is known that, on a given field, one part by weight of corn to two parts by weight of straw is the most favourable proportion for the production of grain, then, according to theory, the manuring of the field should not be such as to cause any marked alteration of this relative proportion in the increased produce; that is to say, the several manuring substances should be selected and laid upon the field in such quantity and relative proportion, that the composition of the soil may remain the same as it was before.

It is well known that certain manuring substances are especially favourable to the formation of the herbaceous parts of plants, others to that of seed. Phosphates, as a general rule, increase the grain crop: whilst of gypsum it is well known that where that substance effects an increase in the produce of clover-hay, this increase is always attended with a marked diminution in the produce of seed. The cultivation of potatoes or Jerusalem artichokes tends to reduce the excessive accumulation in the arable surface soil, of substances which promote the formation of the herbaceous parts of plants. Theoretically, therefore, it is not impossible to maintain a certain uniformity of composition in the soil of a field; but this cannot be effected by carrying on the husbandry of an estate by the system of farm-yard manuring. It will hereafter be shown that by the continuous and exclusive use of farm-yard manure, the composition of the soil is found changed after each rotation.

The last point which claims our attention, in reference to the Saxon experiments, is the difference in the permeability of the soil to the dung-constituents in the different localities. The depth to which the alkalies, the ammonia, and the soluble phosphates penetrate, depends of course upon the absorptive power of the soil; now, assuming, for the sake of illustration, the soil of a field to be divided from the top downwards into distinct layers, which are not of course sharply separated from one another, we find that in some localities the dung-constituents stop in the upper layers, whilst in others they penetrate to the deeper layers of the ground. Thus, for instance, in the Cunnersdorf field the clover crop had derived no benefit from the farm-yard manure, being about only 4 per cent. larger than the produce given by the unmanured plot; whereas at Mäusegast the manuring caused an increase of 30 per cent., and at Oberbobritzsch of 200 per cent. This result shows that certain mineral constituents, indispensable for clover, penetrated much deeper into the ground at Mäusegast and Oberbobritzsch than at Cunnersdorf and Kötitz; or, what comes to the same, that, in the two latter places, they were, on their way downwards, retained by the upper layer of the soil. On comparing the crops given by the unmanured plot at Cunnersdorf with those obtained from the unmanured plots in the other localities, we see that the Cunnersdorf field contained nearly as large a store of straw constituents as the fields at Kötitz and Oberbobritzsch, while it was decidedly poorer in the principal grain constituents, namely, in phosphoric acid and, perhaps, also in nitrogen. Hence, with an equal supply of phosphates and ammonia on the three fields, the topmost layer of the ground at Cunnersdorf, being poorer in these constituents, would retain a great deal more of them than that of the other two fields.

The increase in the potato crop and in the produce of oat-grain and straw, on the Cunnersdorf field, clearly indicates that certain dung-constituents made their way to that layer of the soil from which the roots of the oat-plant principally derive their food, which layer, being richer in corn and straw constituents than the arable surface soil, permitted a small proportion of nutritive substances to pass through it and thus reach the clover.

If we compare with this the field at Kötitz, and look

at its extraordinarily scanty crop of oat-grain and straw, we see at once that in the latter field the deeper layers of the soil were much poorer in corn and straw constituents, but that the topmost layer was much richer in corn constituents than the land at Cunnersdorf.

Although the Kötitz field received above 25 per cent. more farm-yard manure than the Cunnersdorf field, yet only a very insignificant portion of that manure found its way down to the clover, as the layer above had retained the substances nutritive to clover, and these had principally served to benefit the oat-plant. The increase in the produce of oat-grain at Kötitz was more than double that obtained from the Cunnersdorf field. At Mäusegast the relations were similar; from the uncommon abundance of corn and straw constituents in the arable surface soil, the absorptive or retentive power of the latter for the dung-constituents in solution was comparatively less, and a considerable proportion of these substances was thus permitted to reach the deepest layers. The uniform rise of the successive crops obtained from the manured field at Oberbobritzsch evidently shows a very uniform distribution of active dung-constituents, such as might be expected in a soil which, though not exactly sandy, yet contained a larger proportion of sand than any of the other experimental fields.

It is easy to see, that by knowing the absorptive power of the arable soil in these several fields, the farmer is enabled to determine beforehand to what depth the nutritive substances supplied in the manure will penetrate into the ground; and it follows, as a matter of course. that he is able to apply with greater effect the mechanical means at his disposal for promoting the distribution of these elements in the soil, in the right places and in the proper manner.

It would answer no good purpose to expatiate still further on this point; my object has been to direct the attention of the farmer to the different facts or phenomena which are presented by his land during the process of cultivation; because a closer observation of each phenomenon will lead him to reflect upon the cause of it. This is the way to obtain an accurate knowledge of the state and condition of the soil.

Observation and reflection are the fundamental conditions of all progress in natural science; and agriculture presents, in this respect, ample room for discoveries. What must be the feelings of happiness and contentment of the man who, by skilfully turning to proper account his intimate knowledge of the peculiarities of his land, has succeeded, without increased application of labour or capital, in gaining from it a permanent increase of produce? For such a result is not only a personal advantage to himself, but a most important benefit conferred upon all mankind.

How paltry and insignificant do all our discoveries and inventions appear, compared to what is in the power of the agriculturist to achieve!

All our advances in arts and sciences are of no avail in increasing the conditions of human existence; and though a small fraction of society may by their means be gainers in material and intellectual enjoyment, the load of misery weighing upon the great mass of the people remains the same. A hungry man cares not for preaching, and a child that is to learn anything at school must not be sent there with an empty stomach.

Every step in advance, however, made by agriculture serves to alleviate the sufferings and troubles of mankind, and to make the human mind susceptible and capable of appreciating the good and the beautiful that art and science present to us. Improvements in agriculture constitute the only solid foundation for further progress in all other branches of knowledge.

We now proceed to consider the changes brought about in the composition of the soil of a given field by cultivation by the system of farm-yard manuring. The cause to which the restoration of the power of production in the soil by farm-yard manure is attributable, is the same in the case of all soils, without exception, however widely the rotations may differ, or whatever be the nature of the crops cultivated upon them.

By the cultivation of cereals, and the removal of the corn-crops, the arable surface soil loses a certain portion of corn-constituents, which must be restored to it by farm-yard manure, if the future crops are to be kept up to the mark of the preceding ones.

This restoration is effected by the cultivation of fodder-plants, such as turnips, clover, grass, &c., on which the cattle on the farm are fed, and the constituents of which are drawn, in large proportion, from the deeper layers of the ground, where the roots of the cereals cannot penetrate.

These fodder plants are consumed either on the field itself, as turnips in England, or in the stalls. A fraction of the nutritive substances contained in these plants remains in the body of the animals fed upon them, while the remainder, ejected in the form of solid or liquid excrements, constitutes farm-yard manure, the principal bulk of which, however, consists of straw which has served for litter.

In Germany animals are not fed upon potatoes themselves, but upon the refuse from the distilleries of potato spirits, which contains all the nutritive substances taken away from the soil in the potato crop, together with the constituents of the barley-malt that have been used in the process of mashing.

Since the whole of the straw taken away in the crops of the preceding rotation is, as a general rule, returned to the arable soil in the shape of farm-yard manure, the field is, at the outset of the new rotation, as rich as before in the conditions for the production of straw; and there exists, under these circumstances, no ground for a diminution of the straw-crops.

With regard to the clover, turnips, potato-waste, &c., upon which the stock on a farm is fed, there remains, as already stated, in the bodies of the horses, cattle, &c., and full-grown animals in general (which no longer materially increase in weight), only a very small fraction of the constituents of the food consumed; but in the young cattle sent to market, in the bodies of the sheep, in the milk and cheese, a portion of these constituents is retained, which is not returned to the soil in the farm-yard manure. The loss of phosphoric acid and potash which the soil sustains by the sale of cattle and of animal products (wool, cheese, &c.), may be estimated at one-tenth of the quantity of these mineral constituents contained in the potatoes, turnips, or clover; and even this estimate is, perhaps, too high. At all events, it is risking no great error to assume that nine-tenths of all the constituents of the clover, potatoes, or turnips, are returned to the field in the farm-yard manure; whence the arable surface soil, after manuring, is richer for the new rotation in the mineral constituents of potatoes, clover, and turnips, than it was before, as the constituents of the two latter plants have been brought up from the deeper layers of the ground.

The far greater portion of the active dung-constituents is retained by the upper layers of the soil, the deeper layers getting back very little of what has been taken from them; the power of the latter, therefore, to produce as large crops of clover or turnips as before is not restored.

The soil constituents which the animals have derived from the turnips, clover, potatoes, &c., and which remain in their bodies, are very nearly identical, in quantity and quality, with those of the cereals; hence the loss sustained by the land may be estimated as equal to the corn-crops sold, *plus* the corn-constituents which the fodder-plants have given up to the animals on the farm.

The restoration of the power of a field to produce a crop of corn as large as the last naturally presupposes that the conditions required for the production of the new crop should remain the same in the very layer of the soil which supplied the preceding crop; in other words, the substances nutritive to corn which were taken away must be fully returned to the arable surface soil.

If farm-yard manure contained only the constituents of straw and potatoes, and nothing else, manuring a field with it could merely restore the productive power of the arable soil for straw and potatoes, but not for corn. Under these circumstances it would remain as rich as before in food elements for straw and potatoes, but would be poorer for corn to the extent of the whole quantity of corn-constituents taken away in the crops.

If farm-yard manure is to restore the former productiveness of a field for corn, it must necessarily contain an amount of corn-constituents corresponding to the loss sustained, that is to say, as much or even more than has been removed.

The amount of the elements of food for corn contained in the farm-yard manure naturally depends upon the sum of these elements which have passed over into manure, from the cattle feeding upon clover or turnips.

Where this supply exceeds the loss sustained, the arable soil is actually made richer in corn-constituents; but in that case it is enriched also in the conditions for an increased produce of straw and tuberous plants. Where, therefore, the farm-yard manure (by the clover or turnip constituents in it) increases the amount of phosphoric acid and nitrogen in the arable soil, it increases, in a much greater proportion, the quantity of potash and lime, and to some extent also that of silicic acid; and since, as already stated, the whole of the straw-constituents removed from the field are brought back to it in that manure, higher crops of corn, straw, and potatoes are the natural result.

This increase of the produce of all cultivated plants drawing their principal food from the arable surface soil, may go on for a very long time, but in all fields it has a certain appointed limit.

The time comes, sooner or later, for every field, when the subsoil (which is to the clover or turnips what the arable surface soil is to the cereals), suffering a continued drain upon its stores of phosphoric acid, potash, lime, magnesia, &c., begins to lose its productive power for clover or turnips; and thus the nutritive substances, taken away from the arable surface soil in the corn crops, are no longer replaced from the store which existed in the deeper layers, and was brought up by the clover or the turnips. But the high returns of corn given by a field do not necessarily decline with the incipient failure of the clover; for where the arable soil of a field has, after every rotation, received from the clover or turnips more corn-constituents than it had lost by the corn-crop, there may be a gradual accumulation of an excess of these elements of food sufficient to conceal altogether from the farmer the true condition of his land. By introducing into his rotation vetches, white-clover, and other fodder-plants that derive their food from the upper layers of the soil, he succeeds in keeping up his live stock, and he indulges in the notion that all things go on in his field just as before, when the clover or the turnips yielded good crops. This is of course simply a delusion, as there is no longer an actual replacement of the loss sustained. His high corn-crops are now gained at the expense of the nutritive substances accumulated in excess in the arable surface soil which are set in motion by the fodder-plants introduced into the rotation, and are uniformly distributed again in the arable soil after each rotation, by means of the farm-yard manure.

His dung-heap may happen to be of larger bulk and extent than formerly, but as there is now no further supply of nutritive substances brought up from the subsoil or the deeper layers by the clover or turnips, the power of the manure to restore the original fertility of the arable soil is continually decreasing With the ultimate consumption of the excess of corn-constituents accumulated in the arable soil, the time comes when the corn-crop begins to diminish, whereas the produce of straw is comparatively higher than before, as the conditions for the formation of straw have been steadily increasing.

Of course, the farmer cannot fail to remark the diminution of his corn-crops, which induces him to have recourse to drainage, to improved tillage, and to the substitution of other cultivated plants, in lieu of clover and turnips. If the subsoil of his fields will permit it, he now includes in his rotation lucerne and sainfoin, whose still longer and more widely spreading roots enable them to reach yet deeper layers of the ground than the red clover; until finally he employs the yellow lupine, which may truly be called the 'hunger-plant.'

A new increase of produce is the result of these 'improvements' in his system of cultivation by farm-yard manuring, which the farmer looks upon as a great advance. A fresh store of nutritive substances, brought up from the deeper layers of the soil, may possibly accumulate again in the arable surface soil; but these deeper layers also will be gradually exhausted, and the accumulated store in the arable surface soil will also be consumed.

This is the natural termination of cultivation by the system of farm-yard manuring.

The fields of the Saxon experiments afford very fair illustrations of the different conditions to which arable land in general is brought, by a pure system of farm-yard manuring.

The field at Cunnersdorf is in the first stage, the Mäusegast field in the second, the fields at Kötitz and Oberbobritzsch in the third stage, of cultivation by farm-yard manuring, to which we have referred.

At Cunnersdorf the arable soil exhausted by the preceding cultivation becomes with every new rotation richer in the conditions required for the production of grain; not only does the clover replace the loss sustained by the removal of the corn-crops, but a remarkable excess of all nutritive substances will gradually accumulate in the arable soil; and, after a series of years, with the same system of cultivation by farm-yard manuring, the field will be brought to the condition of the land at Mäusegast; which means, that the arable soil will acquire a high productive power for corn and other crops, while the produce of clover will decrease. The fields at Kötitz and Oberbobritzsch most probably were in former times in the same condition as the Mäusegast field is at present; not that they ever yielded crops as large as the latter gives, but merely that the unmanured plots have, at some antecedent period, given better crops than in the year 1851. Without an additional supply of mineral elements derived from meadows or other fields not included in the rotation, the produce must go on continually decreasing, as the supply of mineral constituents brought up by the clover from the subsoil, in these two places, is far from sufficient to make up for what is taken away in the corn-crops.

In the following calculation it has been assumed that of the crops obtained, rye and oats were actually removed, and of potatoes and clover one-tenth was carried away in the form of cattle.*

Cunnersdorf.

	lbs.	Phosphoric acid. lbs.	Potash. lbs.
The arable soil lost by removal of	1176 rye-grain..	10·2	5·5
" "	2019 oats......	15·3	7·7
" "	$\frac{1}{10}$ potato crop	2·3	1·1†
" "	$\frac{1}{10}$ clover crop	4·0	2·0†
Total loss		31·8	16·3
The arable soil had returned to it, in $\frac{9}{10}$ of 9144 lbs. of clover-hay		36·18	95·5
Balance in excess		4·38	79·2

The arable soil of the Cunnersdorf field received, accordingly, in the farm-yard manure, more phosphoric acid and more potash than had been carried off by the corn-crops.

In this calculation, it is a question of no importance how much of the rye or oats was carried off. More

* The amount of phosphoric acid and potash is estimated in the calculation as follows:—

	Rye Corn.	Rye Straw.	Oats Corn.	Oats Straw.	Potatoes.	Clover-hay.
Phosphoric acid.......	0·864	0·12	0·75	0·12	0·14	0·44
Potash	0·47	0·52	0·38	0·94	0·58	1·16

† The quantity of potash is calculated here upon the proportion of phosphoric acid in corn, one part by weight of potash to two parts by weight of phosphoric acid.

than the field produced could not be carried away, and if less were removed the only effect would be that phosphoric acid and potash would accumulate all the more in the field.

Mäusegast.

	Phosphoric acid. lbs.	Potash. lbs.
The arable soil lost by the rye-grain, barley-grain, $\frac{1}{10}$ potatoes, $\frac{1}{10}$ clover	35·4	18·1
The arable soil received in $\frac{9}{10}$ of the clover crop	22·0	62·0
Loss	13·4	Gain 43·9

Kötitz.

	Phosphoric acid. lbs.	Potash. lbs.
The arable soil lost in the rye, oats, and in the $\frac{1}{10}$ of the potatoes and clover	26·4	12·7
It received in the clover	8·5	11·0
Loss	17·9	1·7

The calculation is about the same for the field at Oberbobritzsch as for Kötitz. While the arable soil at Mäusegast, in consequence of the large clover crops produced by it, still continues to gain in potash, the corn-crops are gradually reducing the rich store of potash in the Kötitz field.

These three fields show the effect of a pure system of farm-yard manuring, from which is excluded all supply of manure extraneous to the farm itself.

An additional supply of fodder purchased from other farms, or hay grown on natural meadows, answers the same purpose as an additional supply of manure.

It is self-evident that we cannot give more farm-yard manure to a field than it produces, unless we take the constituents of the manure from some other field, which in that case must lose just as much as the former field gains.

If we direct our attention to manured fields, we find that they give larger corn-crops, and in many cases also larger clover or turnip-crops; the arable soil losing more by the removal of the corn-crop, and receiving more back by the increased produce of farm-yard manure, still the ultimate results remain the same.

In the system of cultivating by rotation of crops, it is found that, for a long time, the arable soil grows with each period of rotation very much richer than it is by nature, in potash as well as in lime, magnesia (the principal constituents of clover and turnips), and in silicic acid.

These substances are the principal conditions for the formation of roots and leaves; their accumulation in the soil tends to make the ground rank and prone to grow weeds,* as the farmer says, an evil which arises as a necessary consequence from cultivation by the system of farm-yard manuring, and which can only be met, as he thinks, by a rotation of crops.

It is generally supposed that the best remedy is the hoe; but though mechanical application may retard the developement of weeds for a time, it cannot effectually prevent them. The hoe has some share in removing them, but not all.

* The most noxious of these weeds are the wild radish (*Raphanus raphanistrum*), the corn cockle (*Agrostemma cithago*), the corn-flower or blue-bottle (*Centaurea cyanus*), the German camomile (*Matricaria chamomilla*), and the corn camomile (*Anthemis arvensis*). All these plants contain, in their ash, as much potash as is found in clover, and 7 to 18 per cent. of chloride of potassium, a salt which forms one of the principal constituents of the urine of animals, and which is brought to the field in the farm-yard manure.

	II. Matric. cham.	I. Matric. cham.	Anthemis arvensis.	Centaurea cyanus.	Agrostemma cithago.
Per cent. ash	8·51	9·69	9·66	7·32	13·20
The ash contains:					
Potash	25.49	32·386	30·57	36·536	22·86
Chloride of potassium	18·4	14·25	7·15	11·88	7·55
Phosphoric acid . .	5·1	7·80	9·94	6·59	6·64
Phosphate of iron	2·39	2·39	4·77	2·34	1·80

(Rüling, 'Annal. d. Chem. und Pharm.' vol. lvi. p. 122.)

The succession of crops in rotation is always made dependent upon the cereals; the preceding crops are selected of such a kind that their cultivation will not injure, but rather improve, the succeeding corn-crop. The selection of the particular kind, however, is always governed by the condition of the soil.

In a field abounding in stalk and leaf constituents, it is often found useful to have wheat preceded by tobacco or rape, rye by turnips or potatoes, since these plants by drawing from the soil a large amount of leaf and stalk constituents serve to restore a more suitable proportion between the straw and corn constituents for the future cereal crop, and at the same time to diminish, in the arable soil, those conditions which favour the growth of weeds.

The preceding observations relative to the produce given by the Saxon fields, both in the unmanured and manured state, afford, in my opinion, a perfect insight into the nature and results of cultivation by the system of farm-yard manuring. In the condition of these fields in their several stages, we may see reflected the history of agriculture.

In the first period, or on a virgin soil, corn-crop is made to succeed corn-crop, and when the produce begins to fail, the culture is simply transferred to a fresh field. The increasing requirements of the growing population, however, gradually put a check upon this plan, and compel a steady cultivation of the same surface; a system of alternate fallowing is now resorted to, and efforts are made to restore the lost fertility of the soil, by manuring with the produce of the natural meadows. After a time, this expedient begins to fail, and leads to the cultivation of fodder-plants, the subsoil being thus turned to account as an artificial meadow. The cultivation of fodder-plants proceeds, at first, without interruption; after a time, longer and longer intervals are interposed between the clover and turnip crops; finally, the cultivation of fodder-plants comes to an end, and with it the system of cultivation by farm-yard manuring. The ultimate result is the absolute exhaustion of the soil, inasmuch as the means for increasing the produce of the soil gradually pass away from it by this system.

Of course, the progress by which these different stages are reached is extremely slow, and the results are felt only by the third and fourth generation. When there are woods near the arable land, the peasant seeks to turn the fallen leaves to account as manure; he breaks up the natural meadows which are still rich in elements of food for plants, and converts them into arable land; then he proceeds to burn down the forests, and to manure his fields with the ashes. When the gradual exhaustion in the productive power of the land has led to a corresponding decrease in the population, the peasant cultivates his land once every two years as in Catalonia, or once every three years as in Andalusia.*

No intelligent man who contemplates the present state of agriculture with an unbiased mind, can remain in doubt, even for a moment, as to the stage which husbandry has reached in Europe. We find that all countries and regions of the earth where man has omitted to restore to the land the conditions of its continued fertility, after having attained the culminating period of the greatest density of population, fall into a state of barrenness and desolation. Historians are wont to attribute the decay of nations to political events and social causes. These may, indeed, have greatly contributed to the result; but we may well ask whether some far deeper cause, not so easily recognised by historians,

* The Emperor Charles V. gave orders that the meadows recently turned into arable land should be restored to their former condition. Even before the time of Charles V. orders of the same nature had been issued by the first Catholic Kings, and at a still earlier period by Pedro the Cruel of Castile. In the beginning of the fifteenth century, Henrique of Castile prohibited the exportation of cattle, on pain of death; and as early as the commencement of the fourteenth century, King Alonzo Onzeno had issued ordinances for the preservation of meadows and pastures. ('Bilder aus Spanien von Karl Freiherrn von Thienen, Adlerflycht.' Berlin: Dunker, p. 241.) All in vain! for what avails the power of even the mightiest monarchs against the irrepressible action of a law of nature?

has not produced these events in the lives of nations, and whether most of the exterminating wars between different races may not have sprung from the inexorable law of self-preservation? Nations, like men, pass from youth to age, and then die out—so it may appear to the superficial observer; but if we look at the matter a little more closely, we shall find that, as the conditions for the continuance of the human race which nature has placed in the ground are very limited and readily exhausted, the nations that have disappeared from the earth have dug their own graves by not knowing how to preserve these conditions. Nations (like China and Japan) who know how to preserve these conditions of life do not die out.

Not the fertility of the earth, but the duration of that fertility, lies within the power of the human will. In the final result, it comes very much to the same thing, whether a nation gradually declines upon a soil constantly diminishing in fertility, or whether, being a stronger race, it maintains its own existence by exterminating and taking the place of another people upon a land richer in the conditions of life.

It can hardly be ascribed to caprice or chance that the cultivator in the *huertas* of Valencia obtains three crops yearly from the same soil, while in the immediate neighbouring district the ground is tilled only once in three years; or that the Spaniards burned down forests in sheer ignorance, in order to use the ashes to restore the fertility of their fields. (See Appendix G.)

Everyone who is at all acquainted with the natural conditions of agriculture, must perceive that the method of culture practised for centuries in most countries could not but inevitably impoverish and exhaust even the most fruitful lands; can it then be supposed that there will be any exception in the case of cultivated lands in Europe, and that like causes will not produce like effects?

Under these circumstances, is it right or reasonable to pay any attention to the doctrines of superficial wiseacres, who, with their wretched chemical analyses find an inexhaustible supply of nutritive substances in any given soil, even in one which can no longer produce clover, turnips, or potatoes, and yet may be rendered capable of producing these plants by manuring with ashes or lime in the right places?

In the face of the daily experience which shows that the corn-fields, if they are to remain fruitful, must be manured after a short series of years, it is a crime against human society, a sin against the public welfare, to disseminate the doctrine that the fodder-plants, which furnish manure to the corn-fields, will constantly find upon the field the conditions of their own growth, that the law of nature applies to one kind of plant only, and has no bearing upon the other. The teaching of these men has no other result than to keep agriculture in the low position which it now occupies. In England it is a mere mechanical handicraft, and in that country manure is regarded as merely the oil which smoothes the wheels and keeps the machine in motion.

In Germany agriculture is a jaded horse, treated with blows instead of fodder; nowhere is its real beauty and the intellectual aspect of its pursuit recognised. Not merely for its utility, but on account of this very intellectual nature of its pursuit, it stands above all occupations; and its practice procures, to the man who understands the voice of nature, not only all the advantages for which he strives, but also those pleasures which science alone can afford.

In human society, ignorance is undoubtedly the fundamental, and therefore the very greatest evil. The ignorant man, however rich he may be, is not protected from poverty by his wealth; while the poor man, who has knowledge, becomes rich by its means. Unconsciously to the ignorant farmer, all his industry, care, and toil only hasten his ruin; his crops gradually diminish, and at length his children and grandchildren, no wiser than himself, are unable to maintain themselves upon the homestead where they were born; their land passes into the hands of the man who has knowledge; for by knowledge capital and power are acquired, and by these, as a matter of course, the helpless are expelled from the inheritance of their forefathers.

As an animal cannot care for himself, the law of nature takes care of him, and is his master; but not so with man, who, if he understands the intentions of God in his creation, is master of the law of nature, which yields to him a complete and willing obedience. The animal brings into the world his perceptions and instincts, which grow up with his growth, and without any effort of his own; but to man the Creator gave the gift of reason, and this distinguished him from the brutes. This is the divine talent, which he should put out to interest, and of which it is said, 'He that hath, to him shall be given; but from him that hath not, shall be taken away even that which he hath.' It is only the interest procured by means of this 'talent' that gives man power over the forces of the earth.

Error arising from want of knowledge is excusable, for no one adheres to it after recognising its existence; and the struggle between error and dawning truth arises from the natural striving of men for knowledge. In this contest truth must grow stronger, and if error prevails, this only proves that truth has yet to grow, not that error is truth.

At all times the 'better' has always been the enemy of the 'good;' but men do not comprehend for all that why, in so many cases, ignorance is the enemy of reason.

There is no profession which for its successful practice requires a larger extent of knowledge than agriculture, and none in which the actual ignorance is greater.

The farmer who practises the system of rotation, depending exclusively upon the application of farm-yard manure, needs very little observation, nay only to open his eyes, in order to be convinced, by innumerable proofs, that whatever may have been the outlay of labour and industry applied to the production of farm-yard manure, his fields have not been thereby increased in the power of bearing crops.

If farm-yard manure was actually able to render a

field permanently richer in nutritive substances than it is by nature, we might expect that a course of manuring for fifty years would necessarily produce a steady increase in the crops.

Now, if farmers who practise the system of rotation, laying aside all bias and prejudice, would compare their present with their former crops, or with those obtained by their fathers or grandfathers, none of them would be able to say that the crops have increased, and only few that the average has remained the same. Most of them would find, that on the average, the straw-crops have turned out higher, but the corn-crops lower, and proportionately lower than they formerly were higher; and that the surplus money which their parents gained by the former high crops, the result of their improvements, as they supposed, must now be paid out again, to purchase manuring substances, which, as people formerly thought, could be 'produced.' Now, however, they begin to learn that though such substances may be produced for a time, they cannot be reproduced in perpetuity.

In like manner, the farmer whose richer ground has enabled him to carry out the three-field system, and whose rich meadows guarantee a supply of manure, who obtains as abundant harvests and as large a weight of corn as the farmer who adopts the system of rotation, and thus surmises that his management has procured what the ground gives of its own free will, will inevitably discover that his fields may be exhausted of the conditions of their fertility, and that it is quite erroneous to suppose that all the farmer's art consists in converting manure into corn and flesh.

A simple law of nature regulates the permanence of agricultural produce. If the *amount* of produce is in proportion to the surface presented by the sum total of nutritive substances, in the soil, the *permanence* of the crops will depend upon the maintenance of that proportion.

This law of compensation, the replacement of nutritive substances which the crops have carried away from the soil, is the foundation of rational husbandry, and must, above all things, be kept in view by the practical farmer. He may renounce the hope of making his land more fruitful than it is by nature, but he cannot expect to keep his harvests up to their average if he allows the necessary conditions for them to diminish in his land.

All those farmers who cherish the notion that the produce of their fields has not declined, have not hitherto been able to appreciate the force of this law. Assuming that they have an excess of nutritive substances to deal with, they think they may continue drawing upon it, until a failure becomes visible, and then they fancy it will be time enough to talk of compensation.

This view results from want of understanding the nature of their own acts.

There surely can be no doubt that to manure a field which already contains an *excess* of nutritive substances is opposed to a rational system of cultivation; for what end could be gained by increasing the nutritive substances in a field where a portion of the elements already existing cannot, on account of their mass, come into operation?

But how can sensible men talk of excess when they are obliged to use manure in order to keep up their harvests, and when their crops decline if they employ no manure?

The simple fact, say others, that in certain districts, as in Rhenish Bavaria, agriculture has flourished since the time of the Romans, and that the ground there is just as rich, nay, gives higher crops than in other lands, is a proof how little reason there is to fear want or exhaustion by continued culture; for if such a thing were likely, it would make itself manifest there sooner than elsewhere.

But in the cultivated lands of Europe agriculture is at all events still very young, as we know with the greatest certainty from records of the time of Charlemagne. His ordinances respecting the management of his own estates (*capitulare de villis vel curtis imperatoris*), wherein directions are given to the stewards, as also the official reports to the Emperor (*specimen breviarii rerum fiscalium Caroli Magni*), sent in by inspectors expressly appointed to survey those estates, are irrefragable proofs that there was then no agriculture worth the name. Very little is said in the *Capitulare* about the cultivation of corn, with the exception of millet. It is reported in the *Breviarium*, that at Stefanswerth (a domain of the Emperor), comprising 740 acres (*jurnales*) of arable land and meadow, capable of supplying 600 cartloads of hay, the commissioners found no corn in store, but on the other hand a large number of cattle, 27 sickles great and small, and only seven broad hoes, to till 740 acres of land!

Upon another estate were found 80 baskets of last year's spelt, equivalent to 400 lbs. of flour (=1⅓ bushel, or somewhat more than 3 hectolitres), 90 baskets of spelt of the current year, from which 450 lbs. of flour could be made. *On the other hand, there were* 330 *hams!*

The crop or stock upon another domain amounted to 20 baskets of spelt (=100 lbs. of flour) of the preceding year, and 30 baskets of spelt, of which *one* was used for seed.

It is easy to see that in those days the breeding of cattle was the chief object, and that the cultivation of corn occupied a very subordinate position in husbandry.* A deed of the period shortly after Charlemagne says on this point: 'Every year, *three yokes* of land upon an estate' should be ploughed and sown with seed furnished by the lord of the manor. (See 'die Getreide-Arten und das Brod von Freih. von Bibra.' Nuremberg: Schmid. 1860.)

Hence we possess not a single trustworthy proof that any one field in Germany or France (perhaps we may make an exception in favour of Italy) has served for the cultivation of corn from the time of Charlemagne to our own age; and the argument for the inexhausti-

* It is worthy of remark that Charlemagne introduced, upon his estates, the three-field system, with which he had become acquainted in Italy.

bility of land is almost childish, because it assumes that corn may be continuously taken from a field, *without restoring* the conditions of reproduction. A field does not necessarily become unfruitful for corn because it has yielded large corn-crops; but it ceases to yield corn-crops if it does not receive compensation for the corn-constituents which have been removed. This compensation is facilitated by the breeding of cattle, in proportion to the extent to which this is carried, and especially when the cultivator is acquainted with the operation of manure. In the time of Charlemagne this was well known, for the winter-crops were manured with dung, distinguished as cattle-dung (called *gor*) and horse-dung (*dost* or *deist*). Besides, the practice of marling was then common in Germany.

With regard to the special instance of Rhenish Bavaria as proving the inexhaustibility of the soil, I had an opportunity last autumn, at a meeting of the Society of Naturalists at Spires, of making particular inquiries about the actual condition of the neighbourhood. Rhenish Bavaria, from the slopes of the Hardt mountains to the Rhine, comprises a district of great fertility: the region is inhabited by an extremely industrious population, distributed in small towns and villages. Almost every artisan, even to the tailor and shoemaker, possesses a small plot of ground, on which he raises his potatoes and vegetables. The export of corn from this district is never thought of, but on the contrary corn and a large quantity of manure are imported from Mannheim, Heidelberg, and elsewhere. The manuring substances obtained from the houses of the towns and villages are carefully treasured and employed, so that there can be no fear of exhaustion, since the removed nutritive substances are restored to the fields. In spite of all this, in no part of Germany is the want of manure more felt than there. On the highways children are constantly seen with little baskets, following the horses and swine, to gather the manure dropped by those animals. In the year 1849, during the political agitation in the Palatinate, the peasants had no more urgent request for the improvement of their condition to lay before the magistrates, than a petition to be allowed to collect 'forestings,' that is, to carry off the natural manure from the forests for the benefit of their fields. They urged that without this (very pitiful) addition to their manure, the future prospects of agriculture in the Palatinate were endangered. In fact, a great quantity of manure is laid out upon the vineyards and tobacco fields, which give none in return; hence the increasing want.

There can be no doubt that in the earliest periods most of our cultivated fields gave a succession of abundant crops, without manuring, as in the case even now, with many fields in the United States of America. But no fact has ever yet been more clearly established by experience than this, that in the course of a few generations all such fields are found perfectly unsuited for the growth of wheat, tobacco, and cotton, and that they recover their fertility only by manuring.

I know full well that recorded facts have as little weight with ignorant 'practical men' as those of political history with practical statesmen, who also act according to 'circumstances and contingencies,' and are simply led when they fondly believe they lead. Still, the reflecting mind cannot fail to be struck by the circumstance, that it is just in countries where the land is most positively known to have given for above 4000 years, without manuring by the hand of man, an uninterrupted succession of abundant crops, that the full action of the great law of restitution is most clearly seen.

We know, most positively, that the corn-fields in the valley of the Nile and the basin of the Ganges remain permanently fruitful, simply because nature has taken upon herself to restore the lost condition of productiveness to the soil in the mud deposited by the inundation of these rivers which gradually raises the land.

All the fields that are not reached by the river lose their productiveness unless manured. In Egypt, the amount of the crop to be expected is calculated from the height of the water of the Nile; and in the East Indies a famine is the inevitable consequence whenever there happens to be no inundation.

Nature herself, in these striking instances, points out to man the proper course of proceeding for keeping up the productiveness of the land. (See Appendix H.)

The notion of our ignorant practical husbandmen, that the soil contains ample store of the elements of food to enable them to pursue their system of agriculture, is due partly to the excellent quality of the land, but also to their skill in robbing it. The man who attempts to gain money by filing the weight of one gold piece from a thousand, cannot plead, in extenuation, that it is remarked by no one, but if discovered he is punished by the law; for everybody knows that the fraudulent act, repeated a thousand times, would ultimately leave nothing of the gold pieces. A similar law, from which, moreover, there is no escape, punishes the agriculturist who would make us believe that he knows the exact store of available food elements in his land, and how far it will go; and who deceives himself when he fancies he is enriching his field by bestowing on the arable surface soil the matters taken from the deeper layers.

There is another class of agriculturists consisting of men with a small stock of knowledge joined to a limited understanding, who, indeed, fully recognise the law of restitution, but interpret it after their own fashion. They assert and teach that part of the law only, and not the whole, applies to cultivated fields; that certain constituents, unquestionably, must be restored to the soil to keep up its productiveness, but that all the others are found in the earth in inexhaustible quantities. They generally base their opinion upon some unmeaning chemical analysis, and demonstrate to the simple agriculturist (for whom alone such disquisitions are intended) how rich his fields still are in some one or other of the mineral constituents, and for how many hundred thousand crops the store will still suffice; as if it could be of the least use for any one to know what the soil contains, if the amount of the available food elements that serve to produce the crops, which is the really important point, cannot be determined.

With such absurd assertions they absolutely hoodwink our 'practical' farmers, who, but for them, might see clearly into matters, but who appear only too willing to accept any assertion that will only leave them at peace, and save them the trouble of 'thinking.'

I remember a case where a swindler offered to sell to a wealthy gentleman, at a high price, a mine of almost pure oxide of aluminium, after having shown him, from chemical works, that oxide of aluminium was indispensable for the production of the metal aluminium, the market price of which was as much as 4*l.* per pound, and that the ore of the mine offered for sale contained nearly 80 per cent. of that valuable metal. The purchaser was not aware that the ore in question is generally known as 'pipe-clay,' an article of almost nominal value, and that the high price of the metal arises from the many changes through which the oxide has to pass to effect its reduction to the metallic state.

It is generally the same with the great stores of potash in the soil. The alkali in the ground, to answer the intended purpose, must, by the agriculturist's art, be converted first into a certain form, in which, alone, it is available as food for plants; and if he does not understand how to effect this conversion, all the potash in his soil is of no earthly use to him.

The notion that the farmer need only restore to his land certain substances, without troubling himself about the rest, might not be prejudicial if those who entertained it confined the application to their own farms; but, as a matter of instruction to others, it is untrue and quite exceptionable. It is calculated for the low intellectual standard of the practical man, who, if he in any way succeeds, by certain alterations, in his system, or by the use of certain manuring agents in obtaining better results than another, attributes his success to his own sagacity rather than to the superior quality of his land. He does not even know that the other has tried the very same plans as himself, only without attaining so favourable a result. Our ignorant practical husbandman starts upon the assumption that all fields are the same in condition as his own, and that, therefore, the same system which answers on his farm ought to do equally well on every other; that the manure which he finds useful ought to be equally useful to others; that the deficiencies in his field are the same in all other fields; that what he exports from his land, others export from theirs; and what he is called upon to restore to his soil, others are equally called upon to restore to theirs.

Although he knows next to nothing of the condition of his own land, with which it would, indeed, require many years of careful observation to become familiar, and is most profoundly ignorant about the condition of the land in any other part; although he never has troubled himself with reflecting upon the causes of his success in the cultivation of his fields, and is quite aware that the advice of agriculturists from other parts, respecting manuring, rotation of crops, and the general treatment of his own land, is not of the slightest use to him, because, as he has found, it is not at all applicable to his district; yet all this does not prevent him from wanting to instruct *others*, and persuade them that *his system* is the only true one, and that they need only do as he does to obtain equally favourable results.

The foundation of all such views is a total misconception of the nature of the soil, the condition and composition of which present an infinite variety of shades.

The fact that many fields that happen to be rich in silicates, and in lime, potash, and magnesia, are, by the growth of corn upon the common farm-yard manuring system, drained only of phosphoric acid and nitrogen, and that the farmer need only look to the replacement of these matters without troubling his mind about the rest, has already been fully discussed. This fact no one can dispute: but it is utterly inadmissible to apply it to the case of other fields, and to make other farmers believe that *they*, too, need not trouble *their* minds about supplying to *their* land potash, lime, magnesia, or silicic acid, and that salts of ammonia and superphosphate of lime will suffice to restore the productiveness of all exhausted fields.

A farmer may, therefore, be quite justified in considering that his field can never grow poorer in potash because he never takes any from it, or that it actually contains a superabundance of potash since every rotation tends to accumulate in the soil a fresh amount of that ingredient; but it is childish of him to think himself justified by this circumstance in assuring another agriculturist, about whose system of cultivation he knows nothing, that the fields of the latter equally contain a superabundance of potash.

There are millions of acres of fertile land (sand and clay-soil), in which the proportion of lime or magnesia in the soil does not exceed that of phosphoric acid, and where provision must be made for replacing the former as well as the latter. Again, there are millions of acres of fertile land, which, like calcareous soils in general, are exceedingly poor in potash, and become absolutely barren without a proper supply of this ingredient.

There are, on the other hand, millions of acres of fertile fields abounding so richly in nitrogen that any additional supply of that element would be mere waste.

Ashes will not promote the growth of clover on fields abounding in potash, whilst the application of manuring agents containing phosphoric acid will have that effect; on the other hand, ashes will make clover grow on land deficient in potash, where bone-earth proves useless; and a simple supply of lime containing magnesia will often suffice to restore the productiveness for clover where the land is deficient in lime and magnesia.

When a farmer, besides corn and flesh, grows and sells other produce, the nature of the required supply of mineral elements is thereby necessarily altered. In the average potato produce of three hectares of land we take away the seed-constituents of four wheat crops, besides about 600 lbs. of potash, and in the average turnip produce of three hectares the seed-constituents of four wheat-crops, besides about 1000 lbs. of potash. A supply of phosphoric acid alone will not suffice, in this case, to keep up the productiveness of the land.

The grower of commercial plants, such as tobacco, hemp, flax, the vine, &c., must in like manner strictly attend to the law of restitution, which, properly interpreted, does not imply that he should bestow the same anxious care upon the replacement of all constituents alike which have been taken away in the crops. It would, for instance, be the height of absurdity to require the tobacco planter who grows his crops on a lime or marl soil, to replace the lime carried off in the leaves of the plant. But it tells him that not all that goes by the name of manure is useful for his fields, and it shows him the difference between manures: it informs him of the loss inflicted upon the soil by the preceding crop, and the supply required to insure future harvests; it teaches him never to allow himself to be guided in his proceedings by the opinions of persons who do not take the slightest interest in him and his land, but always to act upon his own observations. A careful study of the weeds that spring up spontaneously in his fields may frequently prove more useful in this respect than a heap of hand-books on agriculture.

If after the foregoing statements the condition of the cultivated land in Europe, and the decline towards which agriculture is tending by the prevailing system of farm-yard manuring, should still be a matter of doubt to many persons unacquainted with the natural sciences, and who trust only to definite numbers as palpable facts, that doubt may, perhaps, be removed by statistical data on the corn produce of the land in different parts of Germany, which have been collected partly by order of the government.

For a correct appreciation of the importance of these data in the matter, it is necessary in the first place to understand clearly what is meant by an '*average*' crop. By this term is designated the average produce, expressed in numbers, of a field, or a number of fields, or all the fields of a district or country. The figure which represents it is found by adding together the produce of all the fields for a number of years, and dividing the sum total by the latter. There is accordingly a special average produce for every district, by which the next year's crop is judged. Thus we talk of a full, or a half, or a three-quarter average, as the produce happens to come up to the calculated average, or fall one-half or one quarter below it.

The question as to the actual condition of our corn-fields may therefore be put thus: Has there been any change in the figure which at any previous period expressed the average produce of the land, and in what sense? Is that figure higher now than formerly, or has it remained the same or fallen? If the figure is higher, this is of course a sign of an improved condition of the land; if it remains the same, the condition has undergone no change; and if it is lower, there can be no doubt that the condition of the land in that district has declined.

I select for my purpose the statistical data of the produce of the Hessian Rhine district, one of the most fertile provinces of the Grand Duchy of Hesse, with an excellent wheat soil, and inhabited by a most industrious and generally well educated population. ('Statistische Mittheilungen über Rheinhessen, von F. Dael, DLL.' Mayence: 1849. Flor. Kupferberg.)

These data embrace a period of fifteen years, from 1833 to 1847; they refer accordingly to the time when guano was not yet used as manure in Germany. The use of bone-earth was at that time also still very limited, and hardly worth taking into account.

A produce of eleven grains of wheat to every two grains sown, of five and a half accordingly, was held to be an average crop for the Hessian Rhine district. (20 malters = 14 bushels = 5120 hectolitres per hectare = 2·471 English acres.)

Taking the figure 1 to express an average crop, the amount of produce reaped in the Rhine district of Hesse was:—

1833.	1834.	1835.	1836.	1837.	1838.	1839.	
0·85	0·78	0·88	0·72	0·88	0·73	0·61	
1840.	1841.	1842.	1843.	1844.	1845.	1846.	1847.
1·10	0·40	0·90	0·74	1·02	0·63	0·75	0·88

which gives a mean for the fifteen years of 0·79 of the former average.

The productiveness of the wheat land in the Rhine district of Hesse has therefore declined somewhat more than one-fifth.

I know all that may be urged against the accuracy of these figures severally, and their trustworthiness collectively; but if they contain errors, the impartial observer must see that these must tend to the *plus* as well as to the *minus* side, and that it would be most extraordinary in the presence of *plus* errors that all the estimates should have fallen out on the *minus* side.

There is, however, a very simple, and at the same time infallible and irrefutable, proof of the correctness of the conclusions drawn from these figures, in the fact that the cultivation of wheat is on the decrease, that of rye on the increase, in Rhine Hesse, and that many fields on which wheat was formerly grown are now turned into rye fields.

Properly understood, the change from wheat to rye always argues a deterioration in the quality of the soil; the farmer begins to grow rye in a wheat field only when the latter no longer gives remunerative wheat crops.

In Rhine Hesse, a 4½ fold produce of rye is considered an average crop; a wheat soil, therefore, capable of giving only four-fifths of an average wheat-crop, can produce a full average rye-crop.

Now the average produce of rye in the fifteen years is 0·96, which pretty nearly corresponds with the full average.

For spelt, the mean was 0·79 of the average; for barley, 0·88; for oats, 0·88; for peas, 0·67; for potatoes, on the other hand, 0·98; and for colewort and turnips, 0·85.

The statistical data collected in Prussia and Bavaria, which are most reliable, give the same result; and I have not the slightest doubt that it would hold equally true with France and other countries, England included. The visible gradual deterioration of the arable soil cannot but command the serious attention of all men who take an interest in the public welfare. It is of the utmost importance that we do not deceive ourselves respecting the danger, indicated by these signs, as threatening the future of the populations. An impending evil is not evaded by denying its existence or shutting our eyes to the signs of its approach. It is our duty to examine and appreciate the signs: if the source of the evil is once detected, the first step is thereby taken to remove it for ever.

CHAPTER VI.

GUANO.

Composition compared with that of seeds; small amount of potash in it; its action—Guano and bone-earth, similarity of their active ingredients—Guano acts quicker than bone-earth, or a mixture of the latter and ammoniacal salts; reason of this—Oxalic acid in Peruvian guano; the phosphoric acid rendered soluble by its means—Peruvian guano, its effect on the cultivation of corn—Moist guano loses ammonia—Moistening guano with water acidulated with sulphuric acid; effect—Inactivity of guano in dry and very wet weather—Rapidity of its action as a manure, on what dependent—Comparison of the effect of farm-yard manure and guano; effect produced by mixing the two—Guano on a field rich in ammonia—Increased produce by guano, what it presupposes—Exhaustion of the soil by continuous use of guano—Mixture of guano with gypsum and with sulphuric acid—The Saxon agricultural experiments; their results.

PERUVIAN guano generally contains 33 to 34 per cent. of incombustible, and 66 to 67 per cent. of volatile and combustible ingredients (water and ammonia). The latter consist principally of uric acid, oxalic acid, a brown matter of uncertain composition, and guanine. The uric acid amounts occasionally to as much as 18 per cent., the oxalic acid generally to 8 or 10 per cent. of the weight of the guano. The relation of uric acid to vegetation is not known, but it is hardly likely that this substance can have a perceptible share in the fertilising action of guano. To account for this action, then, we have only the ammonia and the incombustible constituents left to consider. An analysis of two samples of guano, made by Dr. Mayer and Dr. Zoeller, in my own laboratory, showed 100 parts of guano ash to contain :—

Potash	1·56	to	2·03
Lime	34·00	"	37·00
Magnesia	2·56	"	2·00
Phosphoric acid	41·00	"	40·00

If we compare with this the composition of the ashes of various seeds, we see at once that the incombustible constituents of guano do not altogether replace the soil constituents carried off in the seeds.

In 100 parts of seed ash are contained,—

	Wheat.	Peas and beans.	Rape.
Potash	30	40	24
Lime	4	6	10
Magnesia	12	6	10
Phosphoric acid	45	36	36

The principal difference between the ash of guano and that of these seeds lies in the deficiency of potash and magnesia in the former.

Agriculturists are generally agreed about the necessity of potash for vegetation, and that a supply is required by fields poor in that ingredient, or drained of it; but the question as to the importance of magnesia for seed formation has not, as yet, met with the same attention, and special experiments in this direction would be very desirable. The fact that much more magnesia is found in the seeds than in the straw unmistakably shows that it must play a definite part in the formation of the seed, which might, perhaps, be ascertained by a careful examination of seeds of the same variety of plants containing different amounts of magnesia. It is a well-known fact that the seeds of the several species of cereals having the same proportion of nitrogen, do not always contain the same nitrogenous compounds, and it is possible that the nature of the latter may, in the formation of the seeds, be essentially influenced by the presence of lime or of magnesia, so that the differences in the proportions of both of these alkaline earths may have a certain connection with the presence of the soluble nitrogenous compounds (albumen and caseïn), or of the insoluble (gluten or vegetable fibrine). Of course, the quantity of potash and soda present would have to be taken into account in an investigation of the kind. The fertilising action of guano is generally attributed to the ammonia in it, and to the other ingredients rich in nitrogen; but accurate experiments made to elucidate this point, by the General Committee of the Agricultural Society of Bavaria, which we shall hereafter have occasion to mention, have shown that whilst the use of guano was found, in many cases, to increase very considerably the produce of corn and straw of a field, the application of an ammoniacal salt containing an amount of nitrogen corresponding to that in the guano produced no perceptible effect on the crop of the same cereal, grown in the same year, upon another plot of the field, when compared with the produce of a third unmanured plot of the same field.

Though the part which the ammonia in the guano plays, in many cases, in increasing the produce, cannot be questioned; yet it is equally certain, on the other hand, that in many other instances the fertilising action of guano must be attributed principally to its other constituents.

If the ash of guano is compared with calcined bones, or bone-earth, it is found that the difference between the two is not very great; yet an amount of bone-earth containing the same proportion of earthy phosphate as in guano, or even two to four times that quantity, has not the same action as the latter manure. Even a mixture of bone-earth with ammoniacal salts in sufficient proportion to make the amount of nitrogen and phosphoric acid equal to that contained in the guano, though more efficacious than bone-earth alone has still a different action from guano. The great distinction between the two lies in the greater rapidity of the action of the guano in the first year, and often even in the course of a few weeks, whilst in the year after it is barely perceptible; that of the bone-earth, on the other hand, is comparatively slight in the first year, but increases in the following.

The cause of this difference of action is the oxalic acid in Peruvian guano, which often amounts to from 6 to 10 per cent. If guano is subjected to lixiviation, the water dissolves sulphate, phosphate, and oxalate of ammonia, which latter salt crystallises out abundantly upon evaporating the solution. But if the guano is moistened with water, without lixiviating, and is then left to itself, it is found, upon extracting with water portions of the mixture from time to time, that the proportion of the oxalic acid in the solution gradually decreases, whilst that of the phosphoric acid increases. A decomposition takes place in this moistened condition of the guano, through the agency of the sulphate of ammonia, by which the phosphate of lime is converted into oxalate of lime and phosphate of ammonia. Peruvian guano is, in this respect, a very remarkable mixture, which could scarcely have been more ingeniously compounded for the purposes of the nutrition of plants; for the phosphoric acid in it becomes soluble only in a moist soil, through which it then spreads in form of phosphate of potash, phosphate of soda, and phosphate of ammonia.

The action of guano may rather be compared to a mixture of superphosphate of lime, ammonia, and salts of potash, which, indeed, in many cases, is equal to it. On a soil abounding in lime, guano is, however, decidedly more advantageous than superphosphate of lime, since the latter, upon coming in contact with the carbonate of lime in the soil, is at once converted into neutral phosphate of lime, which requires to meet with another solvent at the place of formation to effect its diffusion through the soil, whilst phosphate of ammonia spreads through a lime soil just as if there was no carbonate of lime in it. The phosphate of ammonia formed when guano is moistened with water (PO_5+3NH_4O), loses in the air one-third of the ammonia. It is owing to this circumstance that guano, when quite dry, will keep without alteration; whereas, when it has been fraudulently moistened, to increase the weight, it loses, by keeping, considerably in ammonia.

If guano, just before its application on the field, is moistened with water and a little sulphuric acid, sufficient to give the water a slightly acid reaction, the decomposition now mentioned, which otherwise requires days and weeks, is effected in a few hours.

That guano should not produce much effect in very dry weather needs no explanation, because, without water, no substance will act in the ground; that it should, however, equally fail in very wet weather, is, undoubtedly, owing in part to the fact that the oxalic acid is washed out, as an ammoniacal salt, by the rain water, and that there is, accordingly, a corresponding quantity of phosphoric acid not made soluble. By the above simple and cheap means the injurious influence of wet weather upon guano may be completely guarded against, inasmuch as the water and sulphuric acid ensure the conversion into a soluble form of the whole of the phosphoric acid, which could have been brought in to that condition by the oxalic acid.

The rapidity with which a nutritive substance employed in the shape of manure produces an effect, depends essentially upon the speed with which it spreads through the soil, and this, again, upon its solubility; hence it is easy to understand why guano surpasses, in these respects, many other manures.

As regards certainty of action, guano will not bear comparison with farm-yard manure, which, from its nature, is effective in all cases; for farm-yard manure restores to the land all the soil constituents of the preceding rotations, though not in the same proportions, whereas guano restores only some of them, and cannot, therefore, replace farm-yard manure. As guano, however, contains, with the exception of a certain quantity of potash, the chief constituents (phosphoric acid and ammonia) of the exported corn and flesh, the addition of a certain proportion of guano to farm-yard manure may serve to restore the proper composition of the latter, and, with it, also that of the soil.

Let us suppose, for the purpose of illustration, that a hectare of land has been manured with 800 cwt. of farm-yard manure, containing, according to Voelker's analysis, 272 kilogrammes of phosphate, and that the field has, at the end of the rotation, returned the same quantity of farm-yard manure of the same composition, and has lost by the corn and the animal produce exported, altogether 135 kilogrammes of phosphates; the productive power of this field, in so far as it depends upon the phosphates, would not only remain unaltered, but would even be considerably increased, by adding to the 800 cwt. of farm-yard manure supplied to it at the commencement of a fresh rotation, 400 lbs. of guano (with 34 per cent. of phosphates in it).

	Kilogrammes.	
The farm-yard manure supplied to the land . .	272	of phosphates.
In the produce exported the field lost	135	"
There remained in the arable soil	137	"
In the new rotation was added by the fresh supply of 800 cwt. of farm-yard manure . . .	272	"
By the addition of the 400 lbs. of guano . . .	135	"
Altogether	544	"

At the beginning of the new rotation the arable soil contained, accordingly, twice as much phosphates as at the beginning of the preceding one.

It will thus be seen that, under these circumstances, where a field receives back, in the farm-yard manure, a larger share of phosphate than it has lost in the crops, the action of guano upon it will grow feebler from year to year, until at last it ceases to be appreciable.

But the case is very different as regards the application of guano on fields to which a smaller quantity of phosphates is returned in the farm-yard manure than has been lost in the crops, and that have, for instance, been cultivated for half a century upon the farm-yard manuring system. It has already been explained, that on such fields certain constituents of the fodder plants and of straw, more particularly soluble silicic acid and potash, are continually increasing in the arable soil, whilst by the export of corn and flesh its store of mineral substances is reduced by the quantity contained in the exported matters. The two sets of constituents had jointly produced the crop. By taking away the seed-constituents a corresponding amount of the straw and fodder constituents was, accordingly, rendered ineffective. In fields of this description, manuring with guano not only brings up the amount of produce to the former standard, but frequently even increases it to a surprising extent, when the soil contains a large store of other assimilable food elements, which require only the presence of the guano constituents to make them available for nutrition. In the increased produce thus obtained, there is, of course, carried off, together with the guano constituents, also a part of the store of the other food elements; and upon repeated manurings with guano the fertilising effect of that agent must therefore necessarily become feebler in the same proportion as the quantity of these other food elements decreases in the ground. The fertilising action of all compound manures is rarely dependent upon one constituent alone; and as guano contains, in its ammonia and phosphoric acid, two food elements, which require the presence of each other to be available, manuring with guano insures the action of the phosphoric acid, because the particles of the latter are in immediate contact with ammonia particles, that are at the same time also available to the roots; and in the same way the phosphoric acid insures and increases the action of the ammonia.

In a soil abounding in ammonia, manuring with phosphates alone possessing the same degree of solubility, will produce the same effect as guano.

When ammonia salts fail to produce any effect on a field whilst guano is found to act favourably, there is reason to attribute the beneficial effect of the guano principally to the phosphoric acid in it; but in the reverse case the conclusion would not hold equally good, because the salts of ammonia produce two different kinds of effects; they may, under certain circumstances, considerably increase the amount of produce, and yet the favourable effect may not be positively attributed to the action of ammonia as such (see page 86).

The presence in the soil of a sufficient quantity of potash and silicic acid is always presupposed when guano increases the produce of corn; and on a soil rich in potash and magnesia, the application of guano alone insures a succession of crops of such plants, which, like potatoes, require for their growth chiefly potash and magnesia.

Meadows and corn fields which gave at first large

crops with guano, become at last, by the continued use of this agent, frequently so drained of silicic acid and potash, as to lose for many years their original productiveness. At the same time it cannot be denied that there may be many soils which, for several years, by the aid of guano alone, might be made to produce high cereal crops before this state of exhaustion appears; but it will at last inevitably come, and it will then be very difficult to repair the damage.

In 800 cwt. of farm-yard manure with which a hectare of land is manured in a rotation of crops, the soil receives (according to Voelker's analysis) the same quantity of phosphates and of nitrogen as in 800 kilogrammes (15·7 cwt.) of guano; in other words, there is as much of these two elements of food for plants contained in 1 lb. of the latter agent as in 50 lbs. of farm-yard manure. Guano, therefore, contains these elements in the most concentrated form, and permits the application of them to certain parts of the field more conveniently than by farm-yard manure, as is often advantageously done after putting in the seed. In many places, guano is mixed with gypsum to reduce its overpowerful action. The gypsum divides the guano particles and causes them to be more equally distributed over the field; but there is no real diminution of the chemical action of the ammoniacal salts; the gypsum decomposes the oxalate and the phosphate of ammonia into sulphate of ammonia and phosphate and oxalate of lime. The phosphate of lime formed in this way is in a state of infinitely fine division, in which it is most suitable for the roots of plants; however, a small portion only of the phosphoric acid is converted into this state, and with the removal of the oxalic acid, ceases, also, the beneficial influence which the latter exercises in promoting the diffusion of the phosphoric acid.

It will, therefore, be found much more effective to moisten the guano with water to which a little sulphuric acid has been added, and to mix it, after twenty-four hours, with saw-dust, turf-dust, or mould, instead of gypsum, and to strew this mixture over the surface of the field. The rain water dissolves out the phosphat of ammonia, which slowly sinks into the ground, and all parts of the soil with which the solution comes in contact are enriched at the same time with phosphoric acid and ammonia. If to the saw-dust, turf-dust, &c., gypsum is added, it decomposes with the phosphate of ammonia into very finely-divided phosphate of lime and sulphate of ammonia, which are separated by the rain water; the soluble sulphate of ammonia penetrating deeper into the ground and carrying down with it a small quantity of the phosphate of lime, whilst the main bulk of the latter is left on the top.

On land poor in potash, the addition of wood ashes to the guano, moistened with water and sulphuric acid, will be found beneficial, as the carbonate of potash decomposes with the phosphate of ammonia into carbonate of ammonia and phosphate of potash, and the potash does not interfere with the phosphoric acid penetrating into the soil.

The results obtained, in the Saxon experiments, by manuring with guano, afford a clear insight into all the peculiarities observed in the action of this manuring agent.

If we compare the produce severally obtained by manuring with guano and with farm-yard manure (see page 186), we are led to the following considerations on the condition of the experimental field:—

Manuring with guano.

	Cunnersdorf.	Mausegast.	Kötitz.	Oberbobritzsch.
	lbs.	lbs.	lbs.	lbs.
Quantity of guano applied	379	411	411	616
1851.				
Rye corn	1941	2693	1605	2391
" straw............	5979	5951	4745	5877
1852.				
Potatoes	17904	17821	19040	13730
1853.				
Oat corn	2041	1740	1188	1792
" straw............	2873	2223	902	2251
1854.				
Clover	9280	6146	1256	5044

Increase of produce above the unmanured plot (see p. 186).

	Cunnersdorf.	Mausegast. (1853, barley instead of oats.)	Kötitz.	Oberbobritzsch.
	lbs.	lbs.	lbs.	lbs.
Amount of nitrogen in the manure....	49·3	53·4	53·4	80·1
Rye corn............	765	455	341	938
" straw............	3028	1369	1732	2862
Potatoes	1237	925	463	3979
Oat corn	22	451	151	264
" straw............	310	383	455	439
Red clover............	136	608	161	4133

In *Cunnersdorf*, the increase of produce obtained in 1851, over the unmanured field, amounted to—

	Corn. lbs.	Straw. lbs.	Ratio. lbs.
By farm-yard manure (180 cwt.) ...	337	1745 =	1 : 5
By guano (379 lbs.)............	765	3028 =	1 : 3·8

The field at Cunnersdorf was naturally rich in those ingredients which we have designated as S*t* (straw) constituents (silicic acid, potash, lime, magnesia, iron), and the increase of these by the farm-yard manure augmented the straw at the expense of the grain crop. The farm-yard manure contained too little of the K (corn) constituents (nitrogen, phosphoric acid).

This explains the powerful action of guano (which contains chiefly K constituents) upon this field; the increase of corn by its means was more than double that obtained from farm-yard manure, and a more suitable proportion was established between the K and S*t* constituents in the ground.

At *Mäusegast* the increase of produce obtained in 1851, above that of the unmanured field, amounted to—

	Corn. lbs.	Straw. lbs.	Ratio. lbs.
By farm-yard manure (194 cwt.) ...	345	736 =	1 : 2·1
By guano (411 lbs.)............	455	1369 =	1 : 3·0

This field was richer in K and S*t* constituents than the

Cunnersdorf field, and contained, already, an excess of *St* constituents. The K constituents supplied in the guano constituted a much smaller fraction of the whole store already present in the field than was the case with the Cunnersdorf field, and their effect tended rather to increase the produce of straw than that of corn.

The application of guano had the effect of producing the same quantity of straw on the Cunnersdorf as on the Mäusegast field (5951 and 5979 lbs.); but the corn reaped from the latter exceeded that obtained from the former by 752 lbs. The Mäusegast field was much richer in K constituents than the Cunnersdorf field.

At *Kötitz* the increase of produce was—

	Corn. lbs.	Straw. lbs.	Ratio. lbs.
By farm-yard manure (229 cwt.) ...	352	1006	= 1 : 2·8
By guano (411 lbs.)	341	1732	= 1 : 5

The effect of guano upon the straw produce was here out of all proportion greater than that of farm-yard manure, whilst the produce of corn was smaller. It is quite evident that *one* constituent acting more powerfully in the direction of the formation of straw was supplied to the field in larger proportion in the guano than in the farm-yard manure. Experiments with superphosphate (excluding ammonia), or with an ammoniacal salt (excluding phosphoric acid), would have shown to which of these two elements the difference in the produce was owing.

At *Oberbobritzsch* the increase of produce was—

	Corn. lbs.	Straw. lbs.	Ratio. lbs.
By farm-yard manure (314 cwt.) ...	452	913	= 1 : 2
By guano (616 lbs.).............	938	2812	= 1 : 3

As the quantity of guano used at Oberbobritzsch was about 50 per cent. more than in the preceding experiments, no comparison as to amount can be made between the produce of this field and that of the others. What is again remarkable here is the similarity of the condition of this and the Mäusegast field; on both, farm-yard manure gave straw and corn in the proportion of 1·2; guano, in the proportion of 1·3. As regards the power of the soluble guano constituents to pass through the soil, we find from these experiments the same conditions existing as with those of farm-yard manure. At Cunnersdorf and Kötitz the whole guano constituents hardly produced any effect upon the clover crop; whilst at Mäusegast and Oberbobritzsch a perceptible increase was the result.

Silicic acid, which gives strength and firmness to stalks and leaves, is not one of the ingredients of guano; hence, after manuring with guano, the tendency of the cereals to lodge, so much dreaded by agriculturists, is observed on many fields poor in silicic acid, whilst on others abounding in this substance it does not occur. On many soils this tendency may be cured by dressing with lime before applying the guano; and in other cases it may be lessened by mixing dung made from straw with the guano.

If we calculate the increase in the produce of cereals, potatoes, and clover, obtained severally in the years 1851 to 1854, from 100 lbs, of guano we find

100 lbs. of guano gave increase of produce.

	Cunnersdorf.	Mausegast.	Kötitz.	Oberbobritzsch.
	lbs	lbs.	lbs.	lbs.
1851 and 1853. Rye and oats...... ...	1088	646	354	731
1852. Potatoes	326	225	112	646
1854. Clover	36	172	39	670

These results show that the same quantity of guano has an equally dissimilar effect upon different fields as farm-yard manure, and that it is quite impossible to draw from the crops obtained any inference as to the quality or quantity of the manuring agent employed to produce them. The field at Mäusegast had received the same amount of guano as the Kötitz field, both, accordingly, the same quantity of nitrogen and phosphoric acid; yet in cereals and potatoes the increase of produce was twice as great, and in clover much greater in the former than in the latter.

How very little the crops will enable us to draw comparisons between the effects of the several constituents of one and the same manuring agent, may be clearly seen from the results of the experiments at Cunnersdorf and Oberbobritzsch.

At Cunnersdorf, 100 lbs. of guano gave an increase of produce in cereals, potatoes, and clover, containing—

	Nitrogen.	Potash.	Phosphoric acid.	Lime.	
	lbs.	lbs.	lbs.	lbs.	
Increase of produce...	9·2	16·1	3·5	3·6	
The guano contained..	13·0	2·0	12·0	12·0	
More in the manure..	3·8	—	8·5	8·4	less in the crops.
Less in the manure ...	—	14·1	—	—	more in the crops.

At Oberbobritzsch, 100 lbs. of guano gave an increase of produce, containing—

	Nitrogen.	Potash.	Phosphoric acid.	Lime.	
	lbs.	lbs.	lbs.	lbs.	
Increase of produce...	23·0	15·5	6·1	16·9	
The guano contained..	13·0	2·0	12·0	12·0	
More in the manure...	—	—	5·9	—	less in the crops.
Less in the manure ...	10·0	13·5	—	4·9	more in the crops.

The difference in the effect produced by the guano on the two fields is most strikingly exhibited by these tables. At Cunnersdorf the produce reaped contained 30 per cent. less, at Oberbobritzsch 77 per cent. more nitrogen than the manure applied.

CHAPTER VII.

POUDRETTE—HUMAN EXCREMENTS.

Poudrette, nature of ; small amount of the food of plants in it—Human excrement its value—Construction of the privies in the barracks at Rastadt—Calculation of the amount of corn produced by the excrement collected ; importance to the neighbourhood—Its effect not impaired by deodorising with sulphate of iron—The excrement of the inhabitants of towns as manure—Its importance.

POUDRETTE, sold as manure, should consist simply of the desiccated excrements of man made into a transportable form. This is not the case, however, as most poudrettes contain, in reality, only a comparatively small proportion of excrementitious matter. To show this, it will suffice to point out that the poudrette of Montfaucon, which is one of the best sorts, contains 28 per cent., that of Dresden from 43 to 56 per cent., that of Frankfort above 50 per cent., of sand. No kind of poudrette is ever met with in commerce containing more than 3 per cent. of phosphoric acid, and the same amount of ammonia. The construction of privies in dwelling-houses (at least, in Germany) does not make it practicable to keep out the sweepings and other rubbish of the house ; besides, when emptying the pits, it is often the practice, after taking out the fluid contents, to throw into the residuary mass some solid porous body, such as brown-coal or turf-dust, to make it drier and more convenient for removal. All additions of the kind, of course, diminish the percentage of effective and available food elements, and increase the costs of transport. The privy pits, moreover, are but rarely water-tight, and permit the greater part of the urine and other fluid contents to leak away, thus causing the loss of a good deal of the most valuable matter, such as the potash salts, and the soluble phosphates. The following statement will show the great value of the excrement of man. In the fortress of Rastadt and in the soldiers' barracks in Baden generally, the privies are so constructed that the seats open, through wide funnels, into casks fixed upon carts. By this means the whole of the excrements, both fluid and solid, are collected without the least loss. When the casks are full, they are replaced by empty ones.*

The food of the soldier, in Baden, consists chiefly of bread, but also of certain daily rations of meat and vegetables. As the body of an adult does not increase in weight, it needs no particular calculation to make out that the collected excrements must contain the ash-constituents of the bread, meat, and vegetables, and also the whole of the nitrogen of the food.

To produce a pound of corn, the soil has to furnish the ash-constituents of that pound of corn; if we supply these ash-constituents to a suitable field, the latter will thereby be enabled to produce, in a number of years, one pound of corn more than it would have done without this additional supply of ash-constituents. The daily ration of a soldier, in Baden, is 2 lbs. of bread; the excrements of the 8000 men of the different garrisons contain accordingly, per day, the ash-constituents and the nitrogen of 16,000 lbs. of bread, which returned to the soil will fully suffice to reproduce the same quantity of corn as had been used, in form of flour, to bake

* The price of a cart is from 100 to 125 florins = £8 6*s.* 8*d.* to £10 8*s.* 4*d.* It will last about five years. The original outlay incurred by the Army administration in Baden, in 1856 and 1857, for the carts and casks amounting to about £370, was speedily repaid out of the proceeds of the manure.

The collective number of the garrisons of Constance, Freiburg, Rastadt, Carlsruhe, Bruchsal, and Mannheim, averages about 8000 men. The receipts for manure sold were in 1852, £285; in 1853, £315; in 1854, £443; 1855, £400; 1856, £668; 1857, £668; 1858, £680; £50 or £60 are to be deducted from these receipts annually for cost of maintenance, repair, &c., of the carts, &c. ('Journ. of the Agric. Soc. of Bavaria,' April 1860. Page 180.)

the 16,000 lbs. of bread. Reckoning 1½ lb. of corn to 2 lbs. of bread, the excrements of the soldiers in the Grand Duchy of Baden give, therefore, annually, the ash-constituents required for the production of 43,760 cwts. of corn.

The peasants about Rastadt and the other garrison towns, having found out at last by experience the powerful fertilising effect of these excrements upon their fields, now pay for every full cask a certain sum (still rising in price every year), which not only has long since repaid the original outlay, besides covering the annual cost of maintenance, repairs, &c., but actually leaves a handsome profit to the department.

The results brought about in these districts are highly interesting. Sandy wastes, more particularly in the vicinity of Rastadt and Carlsruhe, have been turned into smiling corn-fields of great fertility. Assuming, for the sake of illustration, that the peasants had to furnish the whole corn produced by means of this manure, to the military administrations of the several garrison towns, there would thus be established a perfect circulation of these conditions of life, which would provide 8000 men with bread, year after year, without in the least reducing the productiveness of the fields on which the corn is grown, because the conditions required for the production of corn being thus always returned to the soil, would continue to circulate and yet always remain the same.*

What is said here about the corn-constituents applies, of course, equally to the constituents of meat and vegetables, which, returned to the field, will reproduce as much meat and vegetable matter as has been consumed. The same relation that exists between the in-

* When, some years ago, an order was suddenly issued by the authorities of the city of Carlsruhe, to deodorise and disinfect the pits and cesspools with sulphate of iron, before being emptied, the farmers refused at first to pay any longer for the contents, which they argued were by this treatment deprived of their fertilising virtue. Experience has shown that this is not the case, and the disinfected dung commands as high a price now as the article in its pure state did formerly. The dung in the privy carts requires no disinfecting.

habitants of the barracks in Baden and the fields supplying them with bread, exists equally between the inhabitants of towns and the country around. If it were practicable to collect, without the least loss, all the solid and fluid excrements of all the inhabitants of towns, and to return to each farmer the portion arising from the produce originally supplied by him to the town, the productiveness of his land might be maintained almost unimpaired for ages to come, and the existing store of mineral elements in every fertile field would be amply sufficient for the wants of the increasing populations. At any rate, that store is, at present, still sufficient to do so, although the number of farmers who take care to cover by an adequate supply of suitable manures the loss of mineral matters sustained by the land in the crops grown on it, is but small in proportion to the whole agricultural population. However, sooner or later, the time will come when the deficiency in the store of these mineral matters will be important enough in the eyes of those who are, at present, so void of sense as to believe that the great natural law of restoration does not apply to their own fields; and the sins of the fathers, in this respect, will also be visited upon their posterity. In matters of this kind, inveterate evil habits are but too apt to obscure our better judgment. Even the most ignorant peasant is quite aware that the rain falling upon his dung-heap washes away a great many silver dollars, and that it would be much more profitable to him to have on his fields what now poisons the air of his house and the streets of his village; but he looks on unconcerned, and leaves matters to take their course, because they have always gone on in the same way.

CHAPTER VIII.

EARTHY PHOSPHATES.

High agricultural value of phosphates—Phosphates of commerce; selection of the kind to be used dependent on the object in view, and on the nature of the soil—The rapidity and duration of the effect of the neutral and of the soluble phosphate (superphosphate) of lime—The Saxon manuring experiments.

THE earthy phosphates are among the most important agents for restoring the impaired productiveness of land; not that they influence vegetation in a more marked manner than other mineral elements, but because the system of cultivation pursued by the corn and flesh producing farmer tends to remove them from the soil in larger proportion than other constituents.

In choosing among the phosphates of commerce, the farmer should always keep in view the object which he intends to accomplish, as some sorts will answer better for certain purposes than others.

The so-called superphosphates are commonly phosphates to which a certain quantity of sulphuric acid has been added, to convert the insoluble neutral lime salt into a soluble acid salt. When mixed with a salt of ammonia and a salt of potash, they are often called guano or ammoniacal superphosphates. A good superphosphate generally contains from 10 to 12 per cent. of soluble phosphoric acid. On land poor in clay and lime the superphosphates are particularly suitable for supplying the upper layer of the soil with phosphoric acid. Their effect upon the produce of potatoes and of cereals on such fields is equal to that of Peruvian guano. For turnips and rape, which derive advantage from the presence of sulphuric acid, they have a special value. On chalky soils, the free phosphoric and sulphuric acids are immediately neutralised, by which they are deprived of one of their essential properties, viz., their ready diffusibility, which renders them so valuable a manure for other soils.

Among the neutral phosphates bone-dust holds the first rank. When bones are exposed, under high pressure, to the action of steam, they lose their toughness, and swell up into a soft gelatinous mass, which, after drying, may be readily ground to a fine powder. In this form it spreads, with great rapidity, through the soil; it dissolves in water to a small but perceptible extent, without requiring the presence of any other solvent. What dissolves, under these circumstances, in water, is a combination of gelatine with phosphate of lime, which is not decomposed by the arable earth, and therefore penetrates deep into the ground—a property wanting in the superphosphate. In the moist ground, however, the gelatine speedily putrefies, being converted into ammonia compounds, and the phosphate of lime is then retained by the arable earth. Bone-dust is the agent best adapted to supply phosphate of lime to the deeper layers of the arable soil, for which purpose the superphosphates are not suitable. Bone-earth, or bone-ash, is the name applied to bones freed, by calcination, from the glue or gelatinous part. The animal charcoal of sugar refineries belongs to this category. It must be reduced to the finest powder to render it fully available for manuring purposes. To effect its more speedy distribution through the soil, the presence of a decaying organic substance is necessary to supply the carbonic acid required for its solution in rain water. An excellent way is to mix the powder with farm-yard manure and let the mixture ferment. Among the phosphates of commerce, the guano coming from the Baker and Jarvis Islands are distinguished, before others, by their acid reaction and greater solubility. They contain only a small quantity of an azotised substance, no uric acid, and small proportions of nitric acid, potash, magneisa, and ammonia. The Baker guano contains as much as 80 per cent., the Jarvis guano 33 or 34 per cent. of phosphate of lime; the latter having, besides, 44 per cent. of gypsum. In diffusibility, these guanos, when equally finely powdered, approach nearest to bone-dust: their condition also enables the farmer who wishes to accelerate their action, to convert them most readily into superphosphates (100 parts by weight of Baker guano require 20 to 25 per cent. of concentrated, or 30 to 40 per cent. of the lead chamber sulphuric acid).

The influence of these neutral phosphates upon the produce of a field is generally less marked in the first than in the following years, as it takes a certain time to effect their diffusion through the soil. The speedier or slower manifestation of their action upon a field depends, in a great measure, upon the state of fineness of the powder, to which they have been reduced, the greater or less porosity of the soil, the presence in it of decaying matters, and careful tillage; but, under any circumstances, they require a certain store of soluble silicic acid, and of soda and potash in the soil.

The subjoined table giving the produce obtained, in the years 1847–50, by H. Zenker, at Kleinwolmsdorf, in Saxony, shows the difference between guano and bone-dust as regards rapidity and duration of action. In the first year the guano gave the larger produce, which became smaller in each following year; in the first year the crop from the bone-dust was smaller, but in the succeeding years the increase was most remarkable.

	Bone-dust (822 lbs.).		Guano (411 lbs.).	
	Corn.	Straw.	Corn.	Straw.
	lbs.	lbs.	lbs.	lbs.
1847. Winter corn	2798	4831	2951	4711
1848. Barley	2862	3510	2484	3201
1849. Vetches	1591	5697	1095	4450
1850. Winter corn	1351	2768	732	2481

The 411 lbs. guano contained 53, and the total produce 271 lbs. of nitrogen, or very nearly five times more. The bone-dust contained 37 lbs. of nitrogen, whereas in the total produce there were 342 lbs., or nearly nine times more. The bone-dust gave in the crops altogether 71 lbs. of nitrogen more than the guano. Between the quantity of nitrogen in the manure and the amount of the crops reaped, there is, therefore, no connection whatever.

In the Saxon experiments, the plots manured with bone-dust gave the following results:—

Manuring with bone-dust.

	Cunnersdorf.	Kötitz.	Oberbobritzsch.	Mausegast.
	lbs.	lbs.	lbs.	lbs.
Quantity of bone-dust used	823	1233	1644	892
1851.				
Rye corn	1399	1429	2230	1982
" straw	4167	3707	5036	4365
1852.				
Potatoes	18250	19511	11488	19483
1853.				
Oat corn	2346	1108	1718	1405
" straw	3105	1224	1969	1905
1854.				
Clover	10393	2186	7145	5639

Increase of produce over the unmanured field (see p. 186).

	Cunnersdorf.	Kötitz.	Oberbobritzsch.	Mausegast. (1853, barley instead of oats.)
	lbs.	lbs.	lbs.	lbs.
1851.				
Rye corn	227	165	777	—
" straw	1216	694	2021	—
1852.				
Potatoes	1583	934	1737	2587
1853.				
Oat corn	327	—	190	116
" straw	542	—	157	65
1854.				
Clover	1249	1091	6234	101

The field at Kötitz got 50 per cent. more bone-dust than the Cunnersdorf field; yet its produce of all the crops was lower than that of the latter. The field at Oberbobritzsch got, in 1851, twice the quantity of manure that was applied to the Cunnersdorf field; the result was, in the first year, an increase of corn of 250 per cent., and of straw of 66 per cent. more on the former than on the latter. In the third year, however, the increase of produce of oats, both in grain and straw, was considerably larger at Cunnersdorf than at Oberbobritzsch.

The most curious part of the results is the great difference in the increase of the produce of clover on the several fields; from the field at Oberbobritzsch nearly six times as much clover was obtained as from that at Kötitz, although the former had received only one-fourth more bone-dust than the latter.

A glance at the table shows that in the experiments at Cunnersdorf, Kötitz, and Oberbobritzsch, the quantities of bone-dust severally applied as manure were as 1 : 1½ : 2. A comparison of the increase of produce obtained by bone-earth, just as in the case of guano and farm-yard manure, again demonstrates that there is no connection or relation of dependence between the amount of manure and the increase of the crops.

100 *lbs. bone-dust gave increase of produce*—

	Cunnersdorf.	Kötitz.	Oberbobritzsch.
	lbs.	lbs.	lbs.
1851 and 1853. Rye and oats	280·8	40·1	191
1852. Potatoes	192	75	105
1854. Clover	152	96	380

CHAPTER IX.

GROUND RAPE-CAKE.

Nature and composition of; the diffusibility of its constituents in the soil is comparatively great—Its importance as a manuring agent is small—The Saxon agricultural experiments with rape-cake—The inferences from them.

THE residuary mass left by rape-seed after the extraction of the fatty oil from it by the press, contains a large proportion of a matter abounding in nitrogen, which is nearly related to the caseïn in milk. In addition to this substance, it contains the same incombustible or ash-constituents as the ashes of seeds. The rape-seed ash consists of phosphates, and differs but little in composition from the ash of the grain of rye; phosphates of the alkalies and phosphates of magnesia predominate in it. There is no great error made in assuming that in 100 lbs. of rape-cake a field receives the same amount of the incombustible constituents of rye grain as is contained in 250 to 300 lbs. of the latter.

The azotised matter in rape-cake powder is slightly soluble in water, but its solubility increases with incipient putrefaction; hence the nutritive matters contained in it are much more widely diffused in the ground than, for instance, the principal ingredients of guano, ammonia, and phosphoric acid, which are absorbed, as soon as dissolved, by the earth particles that come in contact with them. Whereas with rape-cake powder this takes place only after its azotised matter has been completely decomposed, and its nitrogen converted into ammonia. This decomposition proceeds, however, pretty fast, and the effect of rape-cake makes itself felt, accordingly, in the very first year of its application.

It is owing to this greater diffusibility of its constituents in the earth that rape-cake appears to exercise a somewhat more powerful effect upon vegetation than guano, for instance, with an equal amount of phosphoric acid.

However, rape-cake holds no very important rank as a manure, simply because very few agriculturists are in a position to procure any considerable quantity of it for manuring purposes. Besides, when its great value as an article of food for cattle shall be more universally known and acknowledged, the increasing price will restrict, still more, its use as a manuring agent; the more so since the excrements of animals fed upon rape-cake contain the principal bulk of the constituents to which is due its efficacy as a fertilising agent.

The following results were obtained, in the Saxon experiments, by manuring with ground rape-cake :—

	Cunnersdorf.	Mausegast.	Kötitz.	Oberbobritzsch.
	lbs.	lbs.	lbs.	lbs.
Manure	1614	1855	1849	3288
1851.				
Rye corn	1868	2645	1578	1946
" straw	5699	5998	4218	4475
1852.				
Potatoes	17374	18997	19165	10442
1853.		barley		
Oat corn	2052	1619	1408	1517
" straw	2768	2298	1550	1939
1854.				
Clover	9143	6659	981	2105

Increase of produce over the unmanured field (see p. 186).

	Cunnersdorf.	Mausegast.	Kötitz.	Oberbobritzsch.
	lbs.	lbs.	lbs.	lbs.
Amount of nitrogen in manure	78·9	88·8	89	157·8
1851.				
Rye corn	692	407	314	493
" straw............	2748	1416	1205	1460
1852.				
Potatoes.............	707	2101	588	691
1853.				
Oat corn.............	33	330	69	—
" straw............	205	458	193	127
1854.				
Clover-hay...........	—	1121	—	1194

Here, again, we see, as in the case of farm-yard manure, guano, and bone-dust, that on no one field did the effect of the rape-cake bear any visible proportion or relation to the quantity used.

1000 *lbs. of ground rape-cake gave increase of produce*—

	Cunnersdorf.	Mausegast.	Kötitz.	Oberbobritzsch.
	lbs.	lbs.	lbs.	lbs.
1851. Rye corn and straw...	2130	989	820	594
1853. Oat corn and straw....	147	424	141	39
1852. Potatoes.............	438	1132	318	210
1854. Clover-hay...........	—	604	—	332

These experiments are interesting in reference to the effect of the nitrogen supplied in the manure. A comparison of the increase of produce obtained at Oberbobritzsch, severally by guano and ground rape-cake, gives the following result in this respect:—

Oberbobritzsch.

	611 lbs. guano = 80 lbs. nitrogen and 74 lbs. phosphoric acid.	3288 lbs. ground rape-cake = 157·8 lbs. nitrogen and 39·5 lbs. phosphoric acid.
1851 and 1853. Rye and oats....	4503 lbs.	2069 lbs.
1852. Potatoes................	3979 "	691 "
1854. Clover-hay..............	4133 "	1194 "

The one field at Oberbobritzsch received in the ground rape-cake nearly double the quantity of nitrogen that the other got in the guano, and the difference in the produce of the two is in the highest degree striking.

In the two experiments—

	In the guano.		In the rape-cake.
The nitrogen in the manures was as...........	1	:	2
In the produce it was:			
" cereals, as	2	:	1
" potatoes, as...........................	5·7	:	1
" clover, as	3·4	:	1

The effect of the nitrogen in the guano was, accordingly, in the cereals four times, in the potatoes twelve times, and in the clover seven times, greater than that of the nitrogen in the rape-cake.

Upon comparing the increase of produce with the amount of phosphoric acid in the two manures, we find that this increase appears to bear some proportion, though yet by no means a definite one, to the amount of phosphoric acid severally contained in them.

The general results of the experiments made, in a four years' rotation, on four different fields at Cunnersdorf, Mäusegast, Kötitz, and Oberbobritzsch, may be summed up as follows:—

The 48 harvests from the unmanured plots and from those manured severally with bone-dust, guano, and ground rape-cake, gave in rye grain and straw, in potatoes, in oats grain and straw, and in clover, by manuring with—

	Bone-dust. lbs.	Guano. lbs.	Ground rape-cake. lbs.
Total amount of nitrogen in crops...	1170	1139	1046
Total amount of nitrogen in crops from unmanured plots.........	910	910	910
Increase of nitrogen over the unmanured plots................	260	229	136
The manure contained nitrogen.....	207	236	415
More than in manure..............	53	less 7	less 279

The manure poorest in nitrogen (the bone-dust) thus actually gave the highest, and the one richest in nitrogen (the rape-cake) the lowest, amount of that element in the produce.

To 100 lbs. nitrogen in the manure, there was obtained of that element in the increased produce—

By bone-dust	125 lbs.
" guano	97 "
" rape-cake...........................	32 "

The amount of phosphoric acid in the crops was from—

	Bone-dust. lbs.	Guano. lbs.	Ground rape-cake. lbs.	Unmanured. lbs.
Phosphoric acid............	361	362	338	292
The manure contained......	1102	288	86	—
The fields gained	741	—	—	—
The fields lost.............	—	74	252	292

CHAPTER X.

WOOD-ASH.

The amount of the food of plants in it—Box-wood ash gives only the half of its potash readily to water—Convenience in mixing wood-ash with earth before applying it—Lixiviated ash, its value—Proper mode of applying ashes as a manure.

IT has already been stated that the proportion of potash is very dissimilar in different wood-ashes; those from hard wood being generally richer in that substance than those from soft wood. The ash of beech-wood gives up to water the one-half of the potash in it, in the form of carbonate of potash, the other half remaining in combination with carbonate of lime, in a compound which is only very slowly decomposed by cold water. The ash of pine-wood generally contains, like tobacco-ash, a larger proportion of lime, so that cold water often seems to fail altogether in dissolving any carbonate of potash out of it. However, the continued action of water succeeds always in gradually extracting from all these ashes the whole of the potash; and since they can be easily ploughed deeply in, they are suited better than all other potash compounds to enrich with that alkali the deeper layers of the arable soil. With wood-ashes that part readily with their potash to water, it will be found useful to mix the ash, before applying it, with an earth that absorbs potash, adding so much of the latter that water poured upon the mixture will no longer turn reddened litmus-paper blue. This operation of mixing can best be performed on the field itself.

Wood-ash which has been extracted with water,

such, for instance, as the residue left in preparation of potash, possesses for many fields a high value as a manuring agent, not only on account of the potash always present in it, but also of the phosphate of lime and soluble silicic acid it contains.

As the upper layers of our corn-fields contain already naturally an excess of potash, in proportion to the other food elements, ash-manuring, when confined to the surface soil, rarely exercises a lasting effect; but where the ash is carried down to the proper depth, it affords an excellent means of obtaining permanent crops of clover, turnips, or even potatoes. Intelligent manufacturers of beetroot sugar use with great success the residuary matter from the distillation of their molasses, which contains all the potash-salts of the beetroot, for manuring their fields, to restore to them the potash removed in the beetroot-crops.

CHAPTER XI.

AMMONIA AND NITRIC ACID.

Source of the nitrogen of plants—Amount of ammonia and nitric acid in rain and dew: Bineau, Boussingault, Knop—Quantity of ammonia in the air—Quantity of nitrogenous food brought to the soil yearly by rain and dew; more present in the soil than is removed by the crops—The general reason for decrease of productive power in soils—Classification of manures according to the amount of nitrogen; assimilable and sparingly assimilable nitrogen; the nitrogen theory; only ammonia according to this theory is wanting; resemblance to the humus theory—Manuring experiments with compounds of ammonia by Schattenmann, by Lawes and Gilbert, by the Agricultural Union of Munich, and by Kuhlmann—The efficacy of a manure is not in proportion to its amount of nitrogen: experiments—Large amount of nitrogen in soils; the experiments of Schmid and Pierre; the arable surface soil contains most nitrogen—Form of the ammonia in the soil; Mayer's experiments—Comportment of soil and farm-yard manure with the alkalies—The ineffective nitrogen of the soil made effective by the supply of ash-constituents that are wanting—Progress in agriculture impossible if dependent on a supply of ammoniacal compounds; results of Lawes' experimen twith salts of ammonia—The artificial supply of ammoniacal manures contrasted with the crops produced and the increase of population—Increase of nitrogenous food by natural means; formation of nitrite of ammonia by oxidation in the air according to Schonbeim—Supply of food in excess necessary to produce corn-crops; reasons—How the necessary excess of nitrogenous food for corn may be obtained from natural sources—The supply of nitrogen in farm-yard manure in the Saxon experiments corresponded to the crop of clover-hay—Loss of nitrogen in lime soils by oxidation; utility of a supply of nitrogen to such soils—Effect of nitrogenous food on the aspect of young plants; on potatoes—Empirical and rational systems of agriculture.

FROM the results of a series of most careful observations extending over a number of years made by Bineau in different parts of France on the amount of ammonia and nitric acid in rain-water, it appears that there fell annually upon the area of a hectare (= $2\frac{1}{2}$ acres) 27 kilogrammes (= 59 lbs.) of ammonia (= 22 kilo. = 48 lbs. nitrogen), and 34 kilogrammes (= 75 lbs.) of nitric acid (= 5 kilo. = 11 lbs. nitrogen); altogether, therefore, 27 kilo. or 54 Zollv. lbs. (= 59 lbs. Eng.) of nitrogen.

For an English acre this makes 21·9 Zollv. lbs. (= 24 lbs. Eng.), and for a Saxon acre 30 Zollv. lbs. These numbers nearly coincide with the observations of Boussingault and Knop.

The yearly average quantity of rain falling in various districts, according to the position and elevation of the localities, is very unequal; and investigations have shown that the amount of ammonia and nitric acid contained in rain-water bears an inverse proportion to the quantity of rain. In districts where the rain falls more seldom or less in quantity, the water is richer in these constituents than in more rainy districts. According to Boussingault, dew is richest in ammonia; according to Knop, not richer than rain-water. (See his valuable memoir in the 8 hefte der 'Landw. Versuchstat. in Sachsen.') But plants receive ammonia and nitric acid not merely by means of rain-water derived from the ground and in dew, but also directly from the atmosphere. The experiments of Boussingault ('Annal. de Chem. et de Phys.,' 3 ser. t. liii.) leave no doubt whatever with regard to the constant presence of ammonia in the air. In a kilogramme of the following substances heated to redness, he found these quantities of ammonia, after three days' exposure to the air upon porcelain plates:—

In 1 kilo. quartz-sand	0·60 milligr. ammonia	
" 1 " bone-ash	0·47	"
" 1 " charcoal	2·9	"

Although we can estimate with tolerable certainty the quantity of ammonia and nitric acid which a field annually receives in rain-water, yet the determination of the same in the dew which moistens plants is not practicable. Just as little can we discover how much ammonia or nitric acid is received by plants directly from the air, simultaneously with carbonic acid.

In the elevated plateaus of Central America, where it scarcely ever rains, the cultivated and wild plants receive their nitrogenous food only from the dew or directly from the air; and we may assume, without risk of error, that the plants which grow in the cultivated fields of Europe have as much ammonia and nitric acid furnished to them by the air and the dew, as is conveyed to them in rain-water. A sandy plain, where no plants grow, receives from the rain as much ammonia and nitric acid as a cultivated field; but the latter derives a greater quantity through the plants, and more from the leafy plants, than from those which are poor in leaves. Let us assume that in the Saxon experiments the cereal plants, potatoes, and clover, raised upon the unmanured land, derived the whole of their nitrogen from the ground, and that *nitrogenous food had not been received either from the air or from the dew;* then the profit and loss of the field in nitrogenous nutriment (according to the assumptions made p. 220, that $\frac{1}{10}$ of the nitrogenous constituents in clover and potatoes were carried off in the form of cattle), may be thus represented:—

The field at Cunnersdorf.

		Produced altogether. Nitrogen.	Lost by crop sold. Nitrogen.	Gained by rain. Nitrogen.
	lbs.	lbs.	lbs.	lbs.
1851.				
Rye corn	1176	22·4	22·4	
" straw	2951	10·6	—	
1852.				
Potatoes	16667	69·8	6·9	
1853.				
Oat corn	2019	30·9	30·0	
" straw..........	2563	6·6	—	
1854.				
Clover-hay.........	9144	202·1	20·2	
			79·5	120
At the beginning of the fifth year the field was therefore richer, in nitrogen, by				40·5

The field at Mäusegast.

	Lost by crop sold. Nitrogen	Gained by rain. Nitrogen.
	lbs.	lbs.
1851. Rye	42·7	
1852. Potatoes	7	
1853. Barley	22·2	
1854. Clover-hay	12·2	
	84·1	120
In 1855 the field was richer in nitrogen by . . .		35·9

It is hardly necessary to carry this calculation any further; for all give the same result, viz. that even on the most unfavourable supposition, a field receives back, by the rain alone, actually more, certainly not less, nitrogenous nutriment, than it loses in the ordinary course of agriculture.

This fact may well justify the assertion that a farmer need trouble himself as little about a compensating supply of nitrogen, as of carbon. Both are, in fact, originally constituents of the air, or capable of again becoming air constituents, and are in the circulation of life inseparable from one another.

From the presence of ammonia and nitric acid in rain-water we are led to infer that a source of nitrogen exists, which without the aid of man, supplies plants with this necessary nutriment. With regard to the other nutritive substances, such as phosphoric acid and potash, which of themselves are not movable, this restoration from natural sources does not exist. Hence, we might have supposed, that when inquiry was made as to the causes which, in consequence of cultivation, diminish the productive power of land, the reason of such decrease would first and chiefly have been sought in those nutritive substances which are of themselves immovable, and not in those which possess the power of circulation; especially when it was ascertained that part at least of the latter spontaneously came back to the field every year. But at every stage in the develöpement of a science, preconceived ideas will for a time assert their sway; and such is the case with those notions which ascribe to nitrogen a preëminent importance in the cultivation of land.

In the consideration of a natural phenomenon, and in the investigation of its causes, we cannot tell at first whether it be simple or compound; whether it be due to one or to several causes; hence we are led to attribute the results to those alone which are *first* discovered in operation. No long time ago, people believed that all the conditions of growth lay in the *seed alone;* then they found that *water*, and next that the *air*, had a very decided influence; bye-and-bye they ascribed to certain organic remains in the *ground*, a most important part in the fertility of the soil. When at length they discovered that, among all the substances used for manure, the excrements of animals and the parts and constituents of animals, surpassed all the rest in operative power; when, too, chemical analysis had shown that nitrogen was the chief element in these substances, it is not surprising that nitrogen was then esteemed the sole, and afterwards the principal, agent in manure.

This process of reasoning is in accordance with nature, and cannot be found fault with. At that time, it was not known that the ash-constituents of plants, potash, lime, and phosphoric acid, play as important a part as nitrogen in the vital processes of plants; nay, not even an idea had been formed of the manner in which the nitrogen of nitrogenous compounds operates. Men simply held by the fact that horn, claws, blood, bones, urine, the solid excrements of animals and men, exerted a favourable influence; while woody substances, sawdust and similar materials, had no effect, or as good as none. If in the one case the presence of nitrogen was the reason of activity, so in the other case the want of nitrogen caused the want of activity; in short, by the operation of nitrogen all facts seemed to be harmonised and explained.

If the nitrogenous manures depended for their activity upon the nitrogen which they contained, it followed necessarily that all of them could not possess the same value for the farmer, because they did not all contain the same amount of nitrogen; those which had more of this substance were manifestly more valuable than those which had less. The amount of nitrogen was easily determined by chemical analysis; hence arose the idea to draw up for the benefit of farmers a list of manures with a figure attached to each showing its relative value; those which were most abundant in nitrogen were considered the most valuable, and stood highest in the list.

In this valuation no importance was attached to the form which nitrogen assumed in the various manures, and just as little to the substances which were present along with the nitrogenous compound. In this list it was quite immaterial whether the nitrogenous combination was in the form of gelatine, horn, or albumen; or whether these substances were or were not accompanied by earthy or alkaline phosphates. Dried blood, claws, horn shavings, woollen rags, bones, rape-cake meal, all figured in one and the same list.

As no definite combination was understood by the word 'nitrogen,' it was impossible to prove that the operation of nitrogenous manures bore any proportion to the amount of nitrogen which they contained.

The introduction and application of Peruvian guano and nitrate of soda afforded the so-called nitrogen theory a foundation to rest upon; no manure could be compared with guano for abundance of nitrogen, while it surpassed all others in the rapidity and strength of its action. The powerful effect produced by it coincided entirely with the nitrogen theory; it corresponded with the high amount of nitrogen in the manure, and chemical analysis furnished satisfactory conclusions with regard to the rapidity of its action. The fact that the influence of guano in increasing the crops was generally more rapid than that of other manures containing an equal amount of nitrogen, made it evident that some one of its constituents possessed a peculiar power which was not present in the other manures; and this constituent was supposed to be more conducive than other nitrogenous compounds to the growth of plants.

The discovery of this constituent presented no difficulty. Chemical analysis showed that Peruvian guano was very rich in salts of ammonia, and that one-half of its nitrogen existed in the form of ammonia. But ammonia was already well known as an element of nutrition for plants, and this afforded an easy solution of the rapidity which marked the operation of guano. Peruvian guano accordingly contained in a concentrated state in the ammonia one of the most important nutritive substances for plants, and this nutriment when dispersed in the soil could be directly assimilated by their roots.

From this time forward a distinction was drawn between the various kinds of nitrogenous manures, and 'assimilable' nitrogen was discriminated from that which was termed 'sparingly assimilable.' Assimilable nitrogen was understood to mean ammonia and nitric acid; but the term 'hard of assimilation' was applied to other nitrogenous substances, which could not be made effective until their nitrogen had been converted into ammonia.

The effect of guano in raising large crops of corn was undeniable; hence it was according to theory assumed as incontestable, that its operation depended upon the amount of nitrogen contained in it; it was further considered as certain, that ammonia was the most effective portion of the nitrogen in guano. It followed, therefore, as a matter of course, that the operation of guano could be produced by substituting a corresponding quantity of salts of ammonia; and the partisans of this theory believed that to increase corn crops at pleasure, nothing further was necessary than to procure the requisite quantity of salts of ammonia at a reasonable price. Humus is the only thing wanting; such was the earlier opinion. Now, it is ammonia is the only thing wanting.

This conclusion was an immense step in advance as regards the views of the importance of nitrogen for plants. Instead of attaching no determinate idea to the word 'nitrogen,' the term had now a fixed and definite meaning. That which formerly was called nitrogen was now termed 'ammonia,' an intelligible, ponderable compound separable from all other substances which are likewise constituents of nitrogenous manures, and capable of being used in experiments, in order to test the truth of the theory itself.

If the operation of guano bore any proportion to its nitrogen, then a quantity of ammonia containing an equal amount of nitrogen must produce not only the same, but a much greater effect; for one-half of the nitrogen in guano exists in the form which is difficult of assimilation, whereas the ammonia could be entirely assimilated.

If in any single experiment, the guano produced a powerful effect, and the corresponding quantity of ammonia was inoperative or weaker, this experiment would be amply sufficient to confute the notion which had been attached to nitrogen. For if this notion was correct, the ammonia ought to operate in all cases in which the guano operated, and exactly in the same manner. The oldest experiments in this direction were made by Schattenmann ('Compt. rend.' t. xvii.).

He manured ten plots of a large wheat-field with sal ammoniac and sulphate of ammonia; an equally large plot remained unmanured. Of the manured plots, one received 162 kilogrammes (= 356 lbs. Eng.) per acre; the others received the double, treble, and quadruple quantity of each of these salts.

The salts of ammonia (says Schattenmann, p. 1130) appear to exert a remarkable influence upon wheat; for, only eight days after manuring, the plant assumed a deep dark-green colour, the sure sign of high vegetative power.

The returns obtained by manuring with the salts of ammonia were the following:—

Muriate of Ammonia employed.			Crop.			
			Corn.	Straw.	Less Corn.	More Straw.
	kilo. lbs.	kilo. lbs.	kilo. cwt.	kilo. cwt.	kilo. cwt.	kilo. cwt.
(1) 1 acre .	none		1182=23	2867=56		
(2) 1 " .	162= 356		1138=22	3217=63	44=0·8	348=6·8
(3) 4 " .	324= 712	324= 712				
	486=1069	486=1069	878=17	3171=62	304=6·0	314=6·0
Average of the four						
Sulphate of Ammonia employed.						
	kilo.	kilo.				
(4) 1 acre .	162		1174=23	3078=60	8=0·15	211=4·0
(5) 4 " .	324	324				
	486	486	903=18	3248=63	279=5·3	381=7·5
Average of the four						

It is easy to see that the expectations which had been founded upon the deep dark-green colour were not realised. The salts of ammonia were so far from exerting any influence in augmenting the corn-crop, that they diminished it in every experiment. In the crop of straw there was a small increase.

In these cases the salts of ammonia had not enlarged the corn crop, but had produced the opposite effect from guano, by which corn crops are generally augmented.

These experiments cannot, however, be regarded as decisive proofs against the view of the action of ammonia, because a comparative experiment with guano was not made at the same time and place. It is not impossible, that upon this particular field guano might have produced the same results. Some years later, Lawes and Gilbert published a series of investigations, which seemed to establish the operative power of ammonia, or rather of salts of ammonia. These investigations were intended to show, that the incombustible nutritive substances of wheat were not, of themselves, sufficient to enhance the fertility of a field, but that the crop of corn and straw stood rather in proportion to the supply of ammonia. In fact, that increased crops could be obtained by salts of ammonia *alone*, inasmuch as nitrogenous manures were peculiarly adapted for the cultivation of wheat.

The experiments of Messrs. Lawes and Gilbert are very far, indeed, from proving the conclusions which they wish to draw; they establish rather the fact that these gentlemen have not the slightest notion of what is meant by argument or proof.

They did not attempt to discover whether salts of ammonia alone could produce from one portion of a field continuous larger crops than were yielded by an unmanured portion of the same field.

Neither did they attempt to discover what crops would be yielded by an equal plot of ground by manuring with superphosphate and potash salts during a series of years. But in the first year they supplied a plot of ground for a whole series of years with the constituents of corn and straw, phosphoric acid and silicate of potash (560 lbs. of bone-earth rendered soluble by sulphuric acid, and 220 lbs. of silicate of potash), and manured it, in the following years, with salts of ammonia only, and they would have us to believe that the increased crops obtained under these circumstances were due to the operation of salts of ammonia *alone!*

The imperfect nature of the experiments made by Messrs. Lawes and Gilbert will appear, perhaps, more striking, if the question which they pretend to solve is stated in another form. We will assume that the point to be proved was, that the high additional crops, yielded by a wheat field manured with guano, were due to the operation of the salts of ammonia in the guano, and that its other constituents had no share in the work. If the guano had been lixiviated with water, and two portions of a field had been manured, the one with guano, the other with the *soluble constituents* of an equal quantity of guano, only two cases could occur; the crop of both plots would be either equal or unequal.

If the crops were equal, it would be manifest that the *insoluble constituents* of the guano had no effect: if the crop upon the plot manured with guano was greater, it would be certain that the insoluble constituents (mineral constituents, as Messrs. Lawes and Gilbert would term them) had some share in producing the additional crop. The extent of this share could perhaps be determined, if a third plot were manured with the insoluble constituents, i. e. with the lixiviated residue of an equal quantity of guano.

If an experimentalist, in carrying out his proof, instead of following this method, had, on the contrary, *lixiviated* the guano, and manured a plot of ground in the *first year* with the *insoluble constituents of the guano*, and in the *subsequent years*, with the *soluble constituents*—and if he had maintained that these soluble constituents, in other words, the salts of ammonia in the guano, had *alone* produced the high additional crops, and that these bore a proportion rather to the salts of ammonia than to the incombustible constituents in the guano, we should have good grounds for concluding that he had simply deceived himself; for, in point of fact, the field had been manured, not with salts of ammonia *alone*, but with *all* the constituents of the guano.

What has here been said in reference to guano, which, as before mentioned, has the same effect as a mixture of superphosphate, potash, and salts of ammonia, may be literally applied to the experiments of Lawes and Gilbert.

They manured their field, in the first year, with a quantity of soluble phosphoric acid, lime, and potash, which very nearly corresponds with the amount of these substances in 1750 lbs of guano; and in the subsequent years they applied salts of ammonia. The arable surface soil of the field had, by previous cultivation, been manifestly exhausted of nitrogenous food; and, under these circumstances, the only wonder would have been if the nutritive substances which operate in guano had been able, *without ammonia*, to yield as large a crop as *with ammonia*.

These experiments are worth notice in the history of agriculture, because they show what statements could be laid before farmers, at a time when ignorance of first principles did not yet permit scientific criticism.

With regard to the influence of ammonia and salts of ammonia there was instituted in the years 1857 and 1858, on the part of the General Committee of the Agricultural Society of Bavaria, a series of comparative experiments in the district of Bogenhausen, as to the operation of guano, and various salts of ammonia containing an equal amount of nitrogen, the results of which are decisive.

The experiments were conducted upon a field (a loam) which had gone through the usual rotation, and which, with ordinary farm-yard manure, had borne rye and then oats twice successively. Of eighteen plots in this field, each 1914 square feet in area, four were manured with salts of ammonia, and one with guano, one plot remained unmanured.

As a starting point for estimating the quantity of manure to be employed, it was assumed that 400 lbs. of guano per acre English (= 493 lbs. avoir.) correspond to the full measure of farm-yard manure usually applied. According to this proportion, 20 lbs (= 24⅔ lbs. avoir.) of guano were reckoned for the area in question.

The samples of good Peruvian guano selected were previously analysed, and in 100 parts a quantity of nitrogen was found corresponding to 15·39 of ammonia. As a general rule, only one-half of the nitrogen in guano is present as ammonia; the other half appears as uric acid, guanine, &c., of the operation of which upon the growth of plants little or nothing, as we have before observed, is known. But it was assumed that the nitrogen in these other substances was just as operative as that in the ammonia, and the *quantum* of the various salts of ammonia (which were likewise analysed previously to ascertain exactly their amount of ammonia) was reckoned in accordance with this assumption. Accordingly, for the above 20 lbs. of guano, 1719 grammes

(= 3·75 lbs.) of ammonia were computed as the equivalent; and each of the other four plots received exactly the same quantity of ammonia, in the salt of ammonia employed for manure.

It is clear that if an increased crop was obtained by means of the guano, and if this was due to the amount of its nitrogen, then each of the other four plots, having received the *same quantity of nitrogen*, must necessarily be affected exactly in the same manner as if they, also, had been manured with 20 lbs. of the same guano. The results were as follow:—

Comparative experiments at Bogenhausen with guano and salts of ammonia containing equal quantities of nitrogen.

HARVEST, 1857.—BARLEY.

	grammes. lbs.		Grain. grammes.	Straw. grammes.
Manured with	5880=13	carbonate of ammonia..	6335	16205
"	4200= 9	nitrate "	8470	16730
"	6720=14¾	phosphate "	7280	17920
"	6720=14¾	sulphate "	6912	18287
"	20 lbs.=24⅔	av. guano	17200	33320
Unmanured		....	6825	18375

Although each of the four plots had received the same quantity of nitrogen, still their respective crops did not correspond; on the whole, the crop from the plots manured with salts of ammonia, corn and straw together, was in each case very little higher than that of the unmanured plot; while the plot manured with guano yielded, for the same quantity of nitrogen, 2½ times more corn, and 80 per cent. more straw, than the average crop of the plots manured with salts of ammonia.

In the subsequent year, this experiment was repeated in a similar manner in the same district with winter wheat. The field chosen, and to which six years previously farm-yard manure had been applied, had borne winter rye, then clover, and then oats, for three years. The oat stubble was broken up and then twice ploughed: on the 12th September, 1857, the seed was sown and harrowed in, on one day: immediately after the sowing there was a moderate thunder shower.

The field was divided into seventeen lots, each of 1900 square feet, which were separated from each other by furrows; each was separately sown and harrowed. The quantity of guano used was 18·8 lbs. (= 23·3 lbs. avoir.), and the weight of the salts of ammonia employed was calculated from the amount of nitrogen in the guano, so that, as in the previous experiment, each plot received an exactly equal amount of nitrogen. The results were the following:—

Experiment in Bogenhausen.

RESULT OF HARVEST, 1858.—WINTER-WHEAT.

	Corn. grammes.	Straw. grammes.
Manured with guano, yielded	32986	79160
" sulphate of ammonia (11·8 lbs. Bav.)..	19600	41440
" phosphate " (11·9 " ")..	21520	38940
" carbonate " (10·6 " ")..	25040	57860
" nitrate " (7·1 " ")..	27090	65100
Unmanured	18100	32986

These experiments show in the clearest manner that it is an error to refer the effect of a powerful nitrogenous manure chiefly to the nitrogen which it contains. No doubt it has a share in the operation of these manures, but their energy is not in proportion to the amount of nitrogen in them.

If ammonia or salts of ammonia increase the produce of a field, their effect depends upon the nature of the soil. What we mean here by the nature of the soil is understood by every one; the ammonia can engender in the soil no potash, no phosphoric acid, no silicic acid, no lime; and if these substances, which are indispensable for the developement of the wheat plant, are not found in the soil, the ammonia cannot produce any effect whatever. If, then, in Schattenmann's experiments, and those at Bogenhausen, there were no results from the salts of ammonia, this did not arise from the fact of these salts being in themselves ineffective; but they were inactive, because the conditions of their activity were wanting. Lawes and Gilbert supplied these conditions to their field, and hence ensured activity to the ammoniacal salts they used.

The results obtained by Kuhlmann respecting the effect of salts of ammonia upon meadows are precisely similar. He manured a piece of meadow land with sulphate of ammonia, and obtained a crop of hay larger than the yield of the unmanured plot, because a certain quantity of phosphoric acid, potash, &c. was rendered active, which without the cooperation of salts of ammonia, would not have been the case. On adding phosphate of lime to the salts of ammonia, the activity of the latter was enhanced in an extraordinary degree; he obtained,—

Return of hay, per hectare, 1844.

		Excess above the unmanured plot.
	kilo.	kilo.
(1) By manuring with 250 kilo. sulphate of ammonia ..	5564	1744
(2) " " 333 kilo. sal ammoniac, with phosphate of lime	9906	6086
(3) Unmanured plot	3820	—

Thus, by sulphate of ammonia *alone*, Kuhlmann obtained rather more than half as much hay again as the yield of the unmanured plot; and by adding phosphate of lime he gained almost three times as much.

Those who maintained the theory of the special importance to agriculture of nitrogen in *manure*, formed a similar notion about the cause of fertility in land.

If, in fact, the efficacy of any manure depended on the enrichment of the soil with nitrogen, exhaustion could be explained only by the diminution of the store of nitrogen; and the manure would restore fertility when the nitrogen which had been removed in the harvest was again supplied by it to the field. Accordingly, the unequal fertility of land must be due to the unequal amounts of nitrogen contained in it; and it would follow that the soil richer in nitrogen must be more fruitful than one which contained less of this element.

This theory, too, came to a pitiful end; since that

which was not true for manures could not possibly hold good for land.

Every one who is acquainted with chemical analysis knows that among the constituents of the soil none can be approximately determined with greater accuracy than nitrogen. In an exhausted soil at Weihenstephan and Bogenhausen, nitrogen was determined by the usual method, and calculated to a depth of 10 inches.

The field contained, per hectare,

	Bogenhausen. kilogr.	Weihenstephan. kilogr.
Nitrogen	5145	5801

On both fields summer barley was cultivated in 1857, and the following returns were obtained, per hectare :—

	Bogenhausen. kilogr.	Weihenstephan. kilogr.
Corn	413	1604
Straw	1115	2580
	1528	4184

Thus, the field at Weihenstephan, containing about the same amount of nitrogen, yielded almost four times as much corn, and more than twice as much straw, as the field at Bogenhausen.

In 1858, these experiments were repeated at Weihenstephan with winter wheat, and at Schleissheim with winter rye ; the result was :—

Nitrogen contained to the depth of 10 *inches, per hectare,*

	Schleissheim. kilogr.	Weihenstephan. kilogr.
	2787	5801
	Crop.	
Corn......	115	1699
Straw	282·6	3030
	397·6	4729

The amount of nitrogen in the field at Schleissheim, as compared with that at Weihenstephan, bears the proportion of 1 : 2; whereas the crops are in the proportion of 1 : 14. These facts are fatal to the opinion that there exists any connection between the amount of nitrogen in a soil, and its powers of production; and in truth no one now entertains this belief. For since Kroker in 1846 determined the nitrogen in 22 kinds of soil from various districts, and discovered that even an unfruitful sand contained more than a hundred times, while in arable soils to a depth of 10 inches there were present from 500 to 1000 times, more nitrogen than is necessary for a good crop, similar investigations have been made in all countries, and Kroker's results have been confirmed.

Since that period the fact has been generally admitted, that the great majority of cultivated soils are far richer in nitrogen than in phosphoric acid; and that the relative proportion of nitrogen present, which had been adopted as the standard for calculating the value of manure, was quite inapplicable for estimating the productive power of land.

Hence, between the chemical analysis of manures, and that of the soil, there arose an irreconcilable contradiction. In the chemical laboratory the effective value of a manure could be accurately determined according to the per centage of its nitrogen; but when the farmer had incorporated his manure with the soil, the determination of the per centage of nitrogen in the ground was no longer of any use in estimating its productive power.

This strange circumstance might well have excited suspicion against the theory of the preponderating influence of nitrogen, for which, as already observed, there is not the slightest evidence in point of fact. But instead of this, the advocates of the theory maintained it steadfastly, and endeavoured to explain the behaviour of the soil upon new and still more extraordinary grounds. It had been observed that a very small fraction of the quantity of nitrogen present in the soil, in the form of guano, farm-yard manure, or nitrate of soda, materially increased the crops; whereas, the effect of other manures, which contained nitrogen not in the form of ammonia or nitric acid, was very unequal in respect of time, and, in the case of horn shavings or woollen rags, was extremely slow. This led to the assumption that the nature of nitrogen was as variable in the arable soil as in manures; one portion was supposed to be in the form of ammonia or nitric acid, and this was, properly speaking, the effective part; another portion, on the contrary, existed in some peculiar form which could not exactly be defined, and was quite ineffective.

Hence the productive power of a soil was, according to this view, not in proportion to the entire quantity of nitrogen in it, but could only be measured by the nitric acid and ammonia which it contained. As the advocates of the theory about the effective operation of nitrogen had been accustomed to shirk proving the truth of their doctrine, as a matter of course they did not trouble themselves about adducing any positive facts in support of this extension of it. They believed that they could establish their point in the following way

When a crop contained in corn and straw as much nitrogen as was equivalent to six, four, three, or two per cent. of the whole quantity of nitrogen in the soil, the reason was that there were present in the field six, four, three, or two per cent. of active nitrogen, while the remaining 94, 96, 97, or 98 per cent, were inoperative nitrogen.

The cause of the effect (the amount of active nitrogen in the soil) was consequently inferred from the effect (the amount of nitrogen in the crops). If more of the whole quantity of nitrogen was in an active form, then higher crops would follow; if the crops were lower, the reason was that there was a deficiency of active nitrogen. If in guano or farm-yard manure additional active nitrogen was supplied, the crops would be increased.

By taking a new standard for estimating the productive power of the soil, the former one for the valuation of manure was virtually abandoned. For when efficiency was allowed only to nitric acid and ammonia in the soil, and denied to all other nitrogenous combinations, it was evidently unwarrantable to place those nitrogenous compounds in manures, which were neither ammonia nor nitric acid, in the same class with these two elements of food.

But in the classified estimate of manures, a high place was given to dried blood, horn shavings, gelatine, and the nitrogenous constituents of rape-cake, all substances which contain neither nitric acid nor ammonia. The favourable effect of these manures was, in the majority of cases, undoubted, but still not determinable by analysis. Of two fields, the one manured with rape-cake, the other not, the former yields a larger corn or turnip crop than the latter, but it is not possible to show that there was more ammonia in the one case than in the other. True, it was assumed that the nitrogenous compounds of these manures, the albumen of the blood, the rape-cake, or the gelatine, was gradually converted into ammonia, and so became operative; but it was taken for granted as a matter of course, that the so-called inoperative nitrogenous compounds present in the soil do not possess the power of yielding ammonia, or of being oxydised into nitric acid.

It was well known, indeed, that if one of two fields contained more lime than the other, the one richer in lime, often did not on that account produce more clover. Yet no one thought of assuming that the lime in the richer field existed in a two-fold condition, operative and inoperative, or that the active portion of the lime had caused the difference in the clover crops.

It was also well known that if two fields be manured with the same bone-earth, the one often gave a higher crop than the other, and yet no one thought of assuming that in the second field the inefficiency of the bone-earth was due to the fact that it had passed into a state of inactivity.

It was further known, that the excess of no individual nutritive substance exercised any influence upon the produce of a field; but it was assumed that the case must be different with nitrogen. A surplus of that element, it was surmised, must act, and if it did not, the cause was not ascribed to the field, but to the nature and condition of the nitrogenous compounds.

From this we see that the notion of nitrogen exerting the principal influence in agriculture led to unexampled confusion of thought and to the most baseless and absurd suppositions. None of the advocates of this theory gave themselves the slightest trouble to extract from the ground one of the nitrogenous compounds, which were deemed inoperative, so as to study its nature; but properties were ascribed to them, of which nothing could be known, because the things themselves were not known.

As the advocates of this theory can say nothing about the nature of the nitrogenous compounds present in the ground, they want to make us believe that nothing at all is known about them. But no one, who has an acquaintance with chemistry, has the smallest doubt or uncertainty respecting the origin of nitrogen in the arable soil. It is derived either from the air, whence it is conveyed to the earth in rain or dew; or from organic substances accuumulated from a series of generations of dead and decayed plants, or else from animal remains contained in the earth, or incorporated with it by man in the form of excrements. Animal and human excrements, bodies of animals in the earth, corpses in their coffins, all vanish, with the exception of their incombustible matters, after a series of years; the nitrogen of their constituents is converted into gaseous ammonia, and is distributed in the surrounding soil. The remains of extinct animal life which are embedded, to an enormous extent, in sedimentary strata, or which of themselves constitute whole masses of rock, attest the extraordinary distribution of organic life in the former ages of the earth; and it is the nitrogenous constituents of these animal bodies, passing over into ammonia and nitric acid, which still play an important part in the economy of the vegetable and animal world.

If the smallest doubt could exist on this question, it is completely removed by the investigations of Schmid and Pierre ('Compt. rend.' t. xlix. pp. 711–715).

Schmid examined (see Peters. 'Acad. Bull.' viii. 161) several specimens of Russian black-earth (tscherno-sem) from the Government of Orel, and among them three from the same field, marked by him as 'virgin soil,' of which we may assume that it had never been subject to agricultural operations; the amount of nitrogen in this soil amounted to—

Amount of nitrogen in the tscherno-sem.

Under the turf	0·99	per cent.	nitrogen
4 werschoks (= 7 inches) deeper..	0·45	"	"
Above the subsoil	0·33	"	"

If we assume a cubic decimètre (= 61 cubic in.) of this earth to weigh 1100 grammes (= 2·4 lbs.), then, calculating for the area of a hectare (= 2½ acres), the ground would contain—

	kilo.		cwt.	
1 decimètre (= 4 inches) deep	10890	=	213	nitrogen
1 " " deeper....	4950	=	97	"
1 " " "	3630	=	71	"
30 centimètres (= 11·7 inches) deep..	19470	=	381	"

In examining a soil in the neighbourhood of Caen, Pierre found in it 19620 kilogrammes (= 385 cwt.) of nitrogen distributed, in the following manner, through a hectare to the depth of one mètre (= 3·3 feet.)

	centimètres.		inches.			kilogr.		cwt.
In the first layer of	25	=	10	deep,	the soil contained	8360	=	164
" second "	25—50	=	10—20	"	"	4959	=	97
" third "	50—75	=	20—30	"	"	3479	=	68
" fourth "	75—100	=	30—40	"	"	2816	=	55
						19614	=	384

Thus, according to both investigations, the uppermost layers, or the proper arable soil (about 10 inches deep), were the richest in nitrogen, while in the lower layers the amount decreased.

Such a condition undeniably proves the origin of nitrogen in the arable soil.

If the upper layers which are constantly deprived of nitrogen by cultivation, contain more of this element than the lower, it necessarily follows that the nitrogen must have come from without. The analysis of the most various kinds of soil in many different lands and districts shows that there is scarcely a single fruitful wheat soil which does not contain at least 5000 to 6000 kilogrammes (= 98 to 118 cwt.) of nitrogen per hectare (= $2\frac{1}{2}$ acres) to the depth of 25 centimètres (= 10 inches); and the simplest comparison of the quantity of nitrogen in the soil, with that which is removed in the crops, proves that the latter amounts to a very small fraction, and that the land is exhausted of all other nutritive substances sooner than of nitrogen.

The experiments of Mayer ('Ergeb. landw. u. agric. Versuche.' München. 1ter Heft, s. 129) show that the behaviour of arable soil with respect to alkalies in watery solution affords no conclusion as to the nature of the nitrogenous compounds therein contained. It had been assumed, that all nitrogen in the earth in the form of ammonia could be separated by distillation with caustic alkalies, and that the portion that was not thus separated did not exist as such. Mayer proved the incorrectness of this assumption; he first discovered, that many earths rich in humous constituents when boiled for four hours (which may be considered equivalent to lixiviation for four hours with boiling water) still retained a very considerable quantity of ammonia. The earths employed in these experiments were (1) earth from the hollow trunk of a tree, (2) garden soil rich in organic matters, from the Botanic Garden, (3) strong clay soil from Bogenhausen.

Ammonia.

One million milligrammes (= 2·2 lbs.) retained at the temperature of boiling water:

	milligr.		grs.		milligr.		grs.		milligr.		grs.
(1) Tree soil,	7308	=	112	(2) Garden soil,	4538	=	70	(3) Clay,	1576	=	24

If an arable soil after saturation with ammonia, by

being placed either in a weak solution of pure ammonia. or in a confined space with ammoniacal gas, or over carbonate of ammonia, is then dried and exposed in thin layers in this dry state to the air for fourteen days, all the ammonia not intimately combined in the soil is evolved, and the same result may be produced by constant washing with cold water. Now if soils thus saturated, the ammonia of which has been accurately ascertained, are exposed to distillation with soda lye, it is found that a considerable portion of the absorbed ammonia is not separable in this way. In the following table, A expresses the quantity of ammonia respectively absorbed by various soils at the ordinary temperature of the air; B, the quantity of ammonia retained by the same soils after twelve to fifteen hours' action of soda lye in a water bath.

One million milligrammes (= 2·2 lbs.) of soil from

	Havannah.	Schleissheim.	Bogenhausen.	Clay soil.
	milligr grs.	milligr. grs.	milligr. grs.	milligr. grs.
A Ammonia...	5520 = 85	3900 = 60	3240 = 50	2600 = 40
B " ...	920 = 14	970 = 15	990 = 15	470 = 7

Under these circumstances, it appears that the power of retaining a certain portion of the absorbed ammonia is very unequal; the Havannah earth (a poor lime soil) retains a sixth of the absorbed ammonia, the soil at Schleissheim the fourth, that at Bogenhausen almost a third.*

* We need not be surprised at this peculiar comportment, for it merely proves that part of the ammonia in the earth is contained in an entirely different form from that of a salt. The salts of ammonia are combinations of ammonium, which can be easily decomposed by alkalies, alkaline earths, and metallic oxides, the alkali taking the place of oxide of ammonium, or the ammonium being displaced by some other metal. But we have no reason to believe, that the ammonia, which by physical attraction is fixed in the porous arable soil, yields its place to another body, and is separable by it, if the latter has not a stronger attraction for the soil.

Carbonate of lime, in the cold, produces scarcely any effect upon sulphate of ammonia; but in an arable soil, which contains carbonate of lime, the salt of ammonia is completely decomposed: lime takes the place of the ammonia, the latter however does not become free, but enters into some other combination, upon which lime has no effect.

This explains the reason why an arable soil saturated with ammonia gives back only a portion after being heated with soda lye for several hours; and it is rather, perhaps, the lengthened operation of water at a high temperature, than the chemical attraction of the soda, that gradually separates, in the form of gas, the ammonia fixed by the soil. In this operation there is no perceptible limit, where the evolution of ammonia ceases; for even after twenty-five hours of continuous heating in a water-bath, the fluid which passes off has still an alkaline reaction.

The above arable soils in their natural condition comport themselves with a boiling solution of soda precisely as if they were partially saturated with ammonia. In the following table, A expresses the total quantity of nitrogen in the form of ammonia, which is obtained from various soils at a red heat with soda lime; B, the quantity of ammonia which is separable from them after twelve to twenty-five hours' heating with a solution of soda.

One million milligrammes of earth = (1 kilo. = 2·2 lbs.) from

	Havannah.	Schleissheim.	Bogenhausen.	Clay soil.
	milligr. grs.	milligr. grs.	milligr. grs.	milligr. grs:
A	2640 = 40·6	4880 = 75.0	4060 = 62·5	2850 = 44·0
B	510 = 7·8	1270 = 19·5	850 = 12	830 = 12·7

These numbers lead to some interesting considerations; they show, among other things, that the third, fourth, or fifth part of all the nitrogen contained in the soil is separable in the form of ammonia; and that after twenty-five hours' distillation with a solution of soda, the fluid which passes off has still an alkaline reaction.

As a *soil saturated with ammonia* retains, after five or six hours' heating with a solution of soda, a third, a fourth, or a sixth of the ammonia absorbed by it, and we cannot assert that the retained portion has changed its nature, and is no longer ammonia; so from the comportment of the earth in its natural condition, and under the same circumstances, we cannot conclude that the nitrogen which by distillation cannot be obtained in the form of ammonia, does not, therefore, exist as such in the earth.

Even if the experiments above described do not afford any proof that all the nitrogen in the ground is in the form of ammonia (a portion, besides, is in most cases present as nitric acid), there is, on the other hand, no proof furnished to the contrary.

Strictly speaking, the discussion of the point in question does not depend on this proof; for it is sufficient to show here, that the comportment of the soil with respect to the amount of nitrogen in it is exactly the same as that of farm-yard manure. Only a small portion of the nitrogen in farm-yard manure, is separable by distillation with alkalies; the much larger portion being obtained only by complete decomposition of the substances.

According to Voelker's analysis, 800 cwt. of fresh farm-yard manure contained—

	1854, November. lbs.	1855, April. lbs.
Nitrogen	514	712
Ammonia { free ... 27·2; in salts... 70·4 }	97·6	74·4

If we compare with this the amount of separable ammonia and the total nitrogen in the soil at Schleissheim and Bogenhausen, we have—

	Schleissheim. lbs.	Bogenhausen. lbs.
800 cwt. of arable soil contain nitrogen	321·6	267·2
Present as separable ammonia	101·6	68·0

It is manifest, that when two soils, not particularly rich in nitrogen, contain just as much *ammonia* as an equal weight of farm-yard manure, if we ascribe the effect of the latter merely to the amount of ammonia which it contains, then the unfruitfulness of the field at Schleissheim is entirely inexplicable.

We assume that the entire quantity of nitrogen in farm-yard manure has a definite share in its operation; and as the nitrogenous matters in the arable soil are originally identical with the substances which form the constituents of manures, it is impossible to ascribe to the one an effect which does not equally apply to the other.

There can be no doubt that the nitrogenous compounds in the ground often exert no influence in increasing the crops, while those in the manures undoubtedly produce a favourable effect. Hence the operation of the nitrogenous compounds in the manure must have depended upon causes which the ground did not supply; and it is clear that the same efficacy can be given to the nitrogenous compounds in the soil, if the farmer will take care to bring into play the causes which produced the favourable operation in the manures.

If we consider, for example, the crops yielded (see pp. 148 and 151) by the two fields at Schleissheim in an unmanured condition, and compare them with the quantity of nitrogen in the soil, the result is—

Nitrogen, per hectare (= $2\frac{1}{2}$ *acres*).

	To the depth of 10 inches.	Produce. Corn.	Straw.
In Field 1 (p. 151), 1858 ...	2787 kilo.	115 kilo.	282 kilo.
In Field 2 (p. 148), 1857 ...	4752 "	644 "	1656 "

Those who maintain that the crops depend upon the nitrogen in the soil, would judge the results of these two experiments somewhat in the following way:—

The amount of nitrogen in both fields is as ....100 : 160
The corn crops as ....100 : 560

If the crops are in proportion to the quantity of effective nitrogen in the soil, it follows that the soil of Field 2 contained, not only altogether, but even proportionately, more than Field 1. If the corn crop in Field 1 = 115 kilogrammes corresponded to the fraction of effective nitrogen in the whole amount of nitrogen = 2787 kilogrammes, then Field 2 ought to have yielded 257 kilogrammes of corn, supposing that the relative proportion of active and inactive nitrogen were the same as in Field 1 (for 2787 kilogrammes, nitrogen : 115 kilogrammes, corn = 4752 kilogrammes, nitrogen :

257 kilogrammes, corn). But, in fact, Field 2 yielded two and a half times as much corn; and therefore the amount of active nitrogen in Field 2 was just in the same proportion greater.

This explanation, very simple in itself, is, however, opposed by the fact that both these fields manured in the same year with superphosphate of lime (prepared from phosphorite) (see pp. 148 and 151), gave the following returns:—

Crop, per hectare.

	Corn.		Straw.	
	kilo.	cwt.	kilo.	cwt.
1858. Field 1 manured with superphosphate of lime	654 =	12·8	1341 =	26·5
1857. " 2 " "	1301 =	25·5	3813 =	75·0

Hence, by the application of three nutritive substances, sulphuric acid, phosphoric acid, and lime, without any increase of the quantity of nitrogen in the soil, as much corn was obtained from Field 1, containing 2787 kilogrammes, nitrogen, as from Field 2, containing 4752 kilogrammes. There was then in the former as much effective nitrogen as in the latter, but it was deficient in certain other substances indispensably necessary to produce an action. Its power to become active was first exhibited when these substances were added to the field. In like manner, the favourable influence of superphosphate upon Field 2 was exhibited; for the crop of this plot, when unmanured, did not correspond to the amount of active nitrogen which it contained; but by the addition of superphosphate the crop rose to more than double. And when to the superphosphate upon Field 1, 137 kilogrammes of common salt, and 755 kilogrammes sulphate of soda were added, there was a still greater increase, i. e. there were now 700 kilogrammes of corn, and 1550 kilogrammes of straw, a still greater quantity of apparently inactive nitrogen having been rendered effective.

The intelligent farmer who reflects upon questions of this kind, will be led to the conclusion, that an essential difference may exist between his own practical experience and the theories of the school which seeks to explain them. When practice tells us that farm-yard manure, guano, and bone earth have restored or increased the crops in certain cases, no one can maintain that these are not real facts, or are not trustworthy. But the observations of the practical man extend no further than these facts; he has not actually remarked that the increased crops were produced by the ammonia in the farm-yard manure, or by that in the guano, or by the nitrogen in the nitrate of soda; all this he is led to believe by persons who themselves know nothing about the matter.

It is certainly a most remarkable circumstance, occurring in no other trade or industry, that in most cases the farmer cherishes representations or theories, for the truth of which he has no evidence; nay, he seems even to give up completely the very idea of inquiring into their correctness. It is quite incomprehensible that he should allow himself to be guided and convinced by facts which have not been remarked by himself upon his own ground, but have been observed in altogether different districts, and which must at least remain doubtful as far as their application to his own land is concerned.

If, during the last ten years, only one farmer in a thousand had resolved to institute experiments upon his own land with ammonia or salts of ammonia to test the theory, whether in fact this manure is useful beyond all others in increasing the corn crops, how soon and how easily would an accurate estimate have been formed of its true value by other farmers!

The simple reflection that not one of the substances nutritive to plants does of itself exert any influence upon their growth, and that several other substances must be present, if the first is to prove useful, should have brought him to the conclusion that the case cannot be otherwise with nitrogen; and that the value of a manure cannot be measured by the amount of nitrogen which it contains; for this presupposes that the nitrogen possesses an operative power, *which must manifest itself under all circumstances*, and that the money which the farmer lays out in its purchase will always ensure an adequate return.

Now, when his common sense tells him that such a supposition is impossible, and that he has only to open his eyes to observe by innumerable facts that ammonia is no exception to other nutritive substances, he will of himself come to the conclusion that the inactivity of the great mass of nitrogen in his field is not due to any condition peculiar to itself, which science can neither investigate nor explain, but that it is inactive, just as phosphoric acid, potash, lime, magnesia, silicic acid, and iron, are inactive, when there is wanting in the ground one of the conditions necessary to make them available.

The theory that by far the greater portion of the nitrogen in the ground is incapable of serving for the nutrition of plants, cannot be proved by the fact that the crops do not bear any proportion to the amount of nitrogen in the soil; for were this the case, then all soils must be equally abundant in all other conditions for the growth of plants, and everywhere possess the same geological and mechanical condition. But this assumption is impossible, for on the whole surface of the globe there are not two districts in which the soils are identical in these respects.

This theory must be strenuously opposed, not only because it is false generally, and that it has never yet been proved to be true even in a single case, but still more on account of the pernicious influence which it exercises upon the practice of the farmer. For since it induces him to suppose that it is impossible to give the necessary efficacy to the store of nitrogen in his land, he will never think even of attempting to do so. Being convinced beforehand that he need not try to raise the treasure buried in his field, he never even makes the attempt.

Since the exact observation made in the cultivation of entire countries and divisions of the globe for centuries past, and also well-established facts, make it probable that a source of nitrogenous food exists, which ensures annually to a cultivated field without the husbandman's aid the return of a portion of the nitrogen, and in a rotation the whole amount of that substance which has been taken away in the crops; and further, that the field may be exhausted of every other nutritive substance, however great its store in the ground may be, because they are never spontaneously restored to the soil by nature—whereas this can never happen to nitrogen; then it is contrary to all the rules of logic in any given case, to ascribe without closer examination the exhaustion of a soil above all other things to a loss of nitrogen.

We might suppose that, apart from the suggestions of common sense, the palpable advantage which would accrue to the farmer imperatively demands that he should take all possible pains to verify the correctness of this fact, and to discover how much nitrogenous food is annually restored to him by the atmosphere. For when he knows how far upon the whole he may calculate upon this source, he can easily arrange his system of cultivation to make it most profitable to him. If the atmosphere supplies him with the whole amount of nitrogen which he removes from his field by a rotation, then he can direct his thoughts to the means of keeping his whole farming operations going in the most effectual manner with the store which he annually collects in his manure heap, without spending any money upon nitrogenous food for his plants. If he finds that the atmosphere restores only a portion of that which has been taken away, and he accurately knows what this portion amounts to, then as circumstances require, he can, with judicious economy, supply from other sources what is lacking; or he may so arrange his system of cultivation as to make the supply of nitrogen from natural sources cover what is removed in the crops.

Every advance in an industrial pursuit has a definite standard of value in the price of the products; and no sensible man would call an alteration in the mode of conducting a business by the name of improvement, unless the price of the products covered the cost of production. When the price of guano exceeds a certain limit, so that the crop realised does not bear a proper proportion to the outlay of capital and labour, this very circumstance prevents its application.

From this point of view farmers might long ago have perceived that the question about the necessity of supplying ammonia to increase the crops of corn, includes another question, whether, on the whole, progress in this respect is, or is not, possible in agricultural practice.

A few considerations only are necessary to bring the farmer to the conviction, which I myself entertain, that if increased production depends upon an augmentation of nitrogenous food in the soil, we must at once renounce all idea of improvement. For my own part, I am much more inclined to believe, that progress is only possible and attainable if the farmer restricts himself to that store of nitrogen which he can collect upon his own ground, avoiding as much as possible all purchase of nitrogenous food from other quarters.

On the average, all the experiments of Lawes in England have shown, that *for one pound of salts of ammonia in manures, two pounds of wheat may be reaped.*

These results, we must remember, were obtained from a field in which one acre without manure of any kind was able to yield, for seven years consecutively, 1125 lbs. of corn and 1756 lbs. of straw; and that all the plots manured with salts of ammonia also received phosphate and silicate of potash.*

On an average, Lawes manured his fields with 3 cwt. of salts of ammonia, and thereby he obtained half as much corn again as the unmanured plot yielded.

We will now assume that the extra crop obtained was exclusively due to the salts of ammonia; we will

* On this point Lawes says ('Journal of the Royal Agr. Soc. of Eng.,' v. xiv. p. 282), that for the production of one bushel of wheat (= 64 to 65 pounds, containing 1 pound of nitrogen) which the soil was made to yield above its natural power, 5 pounds of ammonia were requisite (= 16 pounds of sal ammoniac, or 20 pounds of sulphate of ammonia). He adds, however, that in no single experiment did the extra crop obtained correspond to this estimate.

further suppose that all soils are inexhaustible in phosphoric acid, potash, lime, &c.; and consequently, that the continuous application of salts of ammonia would involve no exhaustion of the soil. If we now reckon how much salts of ammonia, by weight, would be necessary for the kingdom of Saxony, in order to obtain half as much corn again as the unmanured land produces, the result is the following:—The kingdom of Saxony comprised, in the year 1843, 1,344,474 acres (1 acre = 1·368 Eng. acre) of arable land, exclusive of vineyards, gardens, and meadows. If we suppose that each acre yields one corn-crop in two years, and that 4 cwt. salts of ammonia had to be applied in the way of manure, the kingdom of Saxony would require annually 2,688,958 cwt. = 134,447 *tons of salts of ammonia.*

Those who possess even a slender acquaintance with chemical manufacture, and know from what raw materials (animal refuse and gas water) salts of ammonia are procured, must easily see that all the manufactories in England, France, and Germany put together, could not produce so much as the fourth part of the salts of ammonia required by comparatively a very small country, in order to increase its products in the manner proposed.

With a similar distribution we can easily calculate how much salts of ammonia would be required for the German provinces of Austria with 11 million jochen (1 joch = 1·422 Eng. acre) of arable land; for Prussia, with 33 million morgen (1 morgen = 0·631 Eng. acre); for Bavaria, with 9 million tagwerk (1 tagwerk = 0·842 Eng. acre); and even if it were possible to quadruple the manufacture of salts of ammonia, this would have no material influence upon the crops.

The cheapest ammonia is conveyed to Europe in Peruvian guano, which, taking a high average, contains 16 per cent.

Peruvian guano is principally used in the cultivated lands of Europe, as in England, France, the Scandinavian countries, Belgium, the Netherlands, Prussia, and the German States, comprising, exclusive of Austria,

120 millions of inhabitants. Now if we suppose that upon these lands for centuries to come 6 million cwt. (= 300,000 tons) of Peruvian guano, containing 360,-000 cwt. of ammonia, were annually applied, and that it was possible, with the means at present at our disposal, by 5 lbs. of ammonia to raise 65 lbs. additional of wheat, or its equivalent value, then the increased crop of corn would just reach so far as to give each individual in the community 2 *lbs. of corn a day for two days in the year.*

If we assume 2 lbs. of corn or its equivalent to be the average amount of nutriment required by an individual, this makes 730 lbs. annually. According to the supposition made above, 36 million pounds of ammonia would produce thirteen times as much = 468 million pounds of corn or its equivalent, whereby 641,-000 individuals could be nourished for a year.

Supposing the population of England and Wales to increase only 1 per cent. annually, this makes 200,000 individuals in one year, and 600,000 in three years. Now the cereals hypothetically raised by help of the ammonia in 6 million cwt. of guano imported from abroad, would suffice but very few years to support the increased population of England and Wales.

And what would be the state of things six or nine years afterwards in England or Europe, if we were actually dependent upon a foreign importation of ammonia, for the support of the increasing population? Could we import 12 million cwt. of guano in six years, or 18 million in nine years?

We know most positively, that in a few years the source of ammonia in guano will be exhausted; that we have no prospect of discovering a new and richer source; that the annual increase of population, not only in England but in all European countries, is more than 1 per cent.; and, finally, that in proportion to the increase in the population in the United States, Hungary, &c., a corresponding diminution must follow in the exportation of corn from those countries. From these considerations the hope of augmenting the crops of a country by the importation of ammonia must appear utterly vain.

In Germany, a pound of wheat costs at present 4 kreutzers (1½*d.*); a pound of sulphate of ammonia, 9 kreutzers (3½*d.*); and if it were possible with a pound of this salt, added to our ordinary manures, to produce 2 pounds more of wheat, then for every outlay of one florin (2*s.*) in money, the German farmer would receive 53 kreutzers (1*s.* 9*d.*) in corn. This relation of outlay to income is evidently well known in practice, for up to this moment salts of ammonia have nowhere come into general use; and though many manufacturers of manure add a certain quantity of ammonia to their productions, this is chiefly to humour the fancy of farmers for this substance; but none of them can tell what advantage results from this addition. This prejudice will soon disappear of itself, when farmers have learned to make a proper use of the nitrogenous food which nature supplies spontaneously to the land without any aid on their part.

The abundant supply of nitrogenous food in the soil, the increase of the same in well-cultivated ground, the examination of rain-water and of the atmosphere, all facts observed in cultivation in general, prove that, even with the highest system of farming, the soil is not exhausted in its store of nitrogenous food, and that consequently there is a circulation of nitrogen, like that of carbon, which presents to the farmer the possibility of increasing his store of active nitrogen in the soil.

The extraordinary effect of superphosphate of lime in augmenting the crops of corn, turnips, and clover, almost without exception, upon all German lands to which these non-azotised manures have been applied; the operation of the newly-introduced Baker and Jarvis guanos* (which contain no ammonia); the action of lime, salts of potash, gypsum, &c., all show without doubt that an accumulation of nitrogenous food has taken place in the soil, the source of which was, until lately, quite obscure.

* From a communication in the 'Official Gazette,' No. 3, of 1st March,

We had reason enough to believe in a partial restoration to the soil of nitrogenous food by air and rain, but that it should be augmented was quite unexplained; because this presupposed that ammonia and nitric acid were produced from the nitrogen of the atmosphere, in evidence of which we had no facts whatever. Very recently this source of the increase of the nitrogenous food of plants was discovered by Schönbein, and the problem was solved in the most unexpected manner.

In his experiments upon oxygen, Schönbein found that the white fume emitted by a piece of moist phosphorus is not, as was previously believed, phosphoric acid, but *nitrate of ammonia.* I myself had an opportunity of seeing this proved at a lecture, illustrated by experiments, which Schönbein delivered at Munich in the summer of 1860. It is probable, as he states, that in this reaction the nitrogen of the atmosphere, by a kind of induction, combines with three equivalents of water, whereby on the one hand nitrous acid, and on the other ammonia, are formed; just as is well known that under the influence of a higher temperature, nitrite of ammonia is decomposed into water and nitrogen gas. The most striking fact is, this salt is formed under circumstances which we should have been led to suppose were precisely those opposed to its formation; but the production of the peroxide of hydrogen (so easily decomposed by heat), during the slow oxidation of æther, which is attended by a perceptible evolution of heat, is a fact not less certain, and hitherto equally unexplained.

The formation of nitrite of ammonia during this slow process of oxidation made it probable that it takes place everywhere on the earth's surface where oxygen enters

1862, for the Agric. Union in Saxony, the following crops per acre were obtained in 1861:—

		Wheat.	
		Corn.	Straw.
3 cwt. Jarvis guano produced	......	2244 lbs.	4273 lbs.
3 " Baker " "	......	2929 "	5022 "
6 " steamed bones "	......	3015 "	4755 "
Unmanured "	......	1955 "	3702 "

into combination; and consequently that the same process, whereby carbon is converted into carbonic acid, forms also an ever-renewing source of nitrogenous food for plants.

Soon afterwards, Kolbe showed ('Annal. d. Chem. u. Pharm.' bd. 119, s. 176) that if a flame of hydrogen gas is allowed to burn in the open neck of a flask containing oxygen, the interior is filled with the red fumes of nitrous acid.*

Further, Boussingault observed that, in the consumption of common illuminating gas, the water in Lenoir's gas machine contained ammonia and nitric acid; and shortly after, Böttger mentioned, in the 'Annual Report of the Physical Society of Frankfort' (meeting of Nov. 2, 1861), that, according to his experiments, not only in the case of hydrogen, but generally when hydro-carbons were burned, a certain quantity of nitrite of ammonia was always formed, together with water and carbonic acid. Almost contemporaneously with this notice, I received from Schönbein a written communication announcing the very same results which he had obtained in the same way, so that no doubt can remain as to the correctness of this fact.

The practical farmer, who is really anxious to improve his method of cultivation, must be led by these undoubted facts to determine upon ascertaining, with the greatest clearness, the effect of nitrogen in his manures. Before he has been convinced that the atmosphere and rain convey the necessary amount of nitrogenous food to his plants, no one could expect him to renounce the employment of ammonia as a manure. When it is asserted that a farmer can give a maximum of fertility to his land without supplying to it any nitrogenous matter, it is not meant that he must renounce the use of farm-yard manure; but the assertion implies the existence of the latter, and is, in fact, based upon it.

For the restoration or augmentation of productive

* The formation of nitrous acid in eudiometrical experiments was already known.

power in exhausted corn-fields, it is absolutely necessary that the arable soil should contain a surplus of all nutritive substances for cereal plants, nitrogenous among others, but no one in greater proportion than the rest. It is assumed that the farmer by a right succession of crops, that is, by a proper proportion between his corn and fodder fields, is always in a position, by carefully husbanding the ammonia in his farm-yard manure and avoiding all unnecessary waste, to provide the arable soil with such a surplus of nitrogenous food as will correspond to the proportion of the other nutritive substances therein stored; and that the atmosphere annually makes up what he removes in his crops.

The nitrogenous food conveyed by the atmosphere and rain, is upon the whole sufficient for his cultivated plants, but not enough for many of them in point of time. In order to give a maximum crop, many plants require, during the period of vegetation, much more than the air and rain afford in that time; and therefore the farmer makes use of fodder plants in order to increase the crops of his corn-fields. The fodder plants, which thrive without rich nitrogenous manure, collect from the ground and condense from the atmosphere, in the form of blood and flesh constituents, the ammonia which is supplied from these sources; and the farmer, in feeding his horses, sheep, and cattle with the turnips, clover, &c., receives, in their solid and fluid excrements, the nitrogen of the fodder in the form of ammonia and products rich in nitrogen; and thus he obtains a supply of nitrogenous manures or nitrogen, which he gives to his corn-fields.

The rule is, that for certain plants, weak in developement of leaf and root, and which have but a short period of vegetation, the farmer must compensate by the *quantity* of manure for the time which is *wanting* for the absorption of the requisite amount of nitrogen from natural sources.

It is easy to see that the accumulation of nitrogenous food by farm-yard manure in the uppermost layers of the ground, so very important for the perfect growth of cereal plants, must chiefly depend upon the successful growth of fodder plants.

The unmanured fields in the Saxon experiments—

	Yielded altogether. Nitrogen.	Lost by sale of crop. Nitrogen.	Received in farm-yard manure. Nitrogen.	Clover crops.
1851–1854.	lbs.	lbs.	lbs.	lbs.
Cunnersdorf............	342·4	78·4	263·6	9144
Mäusegast	279·5	84·1	175·0	5538
Kötitz	160·9	54·8	106·1	1095
Oberbobritzsch........	127·7	57·2	70·5	911

It is easily perceived from this table that the quantities of nitrogen which could be obtained from the field and restored in the form of farm-yard manure, bear a proportion not exact but sufficiently well marked, to the crops of clover produced by the field; and there can be no doubt that the farmer who takes the right way to make his fodder plants thrive, obtains at the same time the means of enriching his arable soil with a surplus of nitrogenous food for his corn-plants.

We do not mean to imply that in every possible case the farmer must renounce the idea of supplying to his land ammonia from other quarters; for soils vary so very much in their nature, that even though we can assert that by far the greater proportion of them may not require a restoration of nitrogenous food, yet this will not hold good for all without exception. In a soil rich in lime and humous materials, in consequence of the process of decay going on, a certain quantity of the ammonia fixed in the earth is converted into nitric acid, which is not retained by the soil, but is conveyed into the lower layers in the form of salts of lime or magnesia. Under certain circumstances, this loss may amount to much more than is compensated by the atmosphere, and for such fields a supply of ammonia will always be useful. The same holds good for certain soils which have not been tilled for many years, and in which, by the operation of the causes above-mentioned, the necessary surplus of nitrogenous food, formerly present, is gradually expended. On recommencing the cultivation of such soils, the employment of nitrogenous manures will at first produce a remarkably beneficial effect. Afterwards, these too require no further supply.

There is one reason which excites in the farmer's mind a prejudice in favour of introgenous manure, and that is the great inequality in the appearance of the young crops, when such manures are applied in comparative experiments. The cereal plants upon fields manured with guano or nitrate of soda are distinguished before others by a deep green colour, and by broader and more numerous leaves; but the harvest is generally far from corresponding to the expectations raised by this promising appearance. Upon a field excessively rich in nitrogenous food, there is a kind of rankness in the early growth like that produced by a hot-bed: the leaves and stalks are watery and weak, in consequence of the want of time in their over-hasty growth to absorb contemporaneously from the soil the necessary quantity of substances, such as silicic acid and lime, capable of communicating to their organs a certain solidity and power of resistance against those external causes which endanger their existence. The stalks fail to acquire the necessary stiffness and strength, and are always liable to be laid, especially on lime soils.

This injurious influence of excess of nitrogenous food is particularly remarkable in the case of the potato plant; for if it grows upon a soil excessively rich in nitrogenous food, and the temperature should suddenly fall and wet weather supervene, the plant is often attacked by the so-called potato disease; while a neighbouring potato field merely manured with ashes shows no trace of it.

Among all the many experiments which have been hitherto made by farmers to improve their land, there is not one instituted for the purpose of ascertaining the actual condition of their soil, or of seeking proofs for the correctness of the notions which they had once adopted. The reason of their indifference about obtaining proofs for their views chiefly consists in this, that the practical man, like the artisan, is guided in his business not by ideas, but by facts. Hence it is quite indifferent to him, whether the theory, or what he dignifies by that name, is correct or not, as he does not regulate his proceedings in accordance with it.

Many thousand farmers, who have not the remotest conception of the nutrition of plants or the composition of manures, apply guano, bone earth, and other manures, to their fields, with fully the same effect and with even the same skill as others who possess such information; nor do the latter derive any manifest advantage from their knowledge, because it is not of the right kind; for example, the chemical analysis of manures is rather calculated for ascertaining their purity, and for determining their price, than as a means for making us acquainted with their effect upon land.

In England bone earth was used and valued as a manure half a century before any idea was formed as to what its operation was due; and when afterwards the erroneous theory was adopted that its effect depended upon the nitrogenous gelatine which it contained, this view did not exert the slightest influence upon its employment.

The farmer manured his field with bone earth, not on account of its nitrogen, but because he wished to have larger crops of corn and fodder, and because experience told him that he could not expect them without bone earth.

An agricultural practice, founded upon a simple acquaintance with facts, without any idea of their nature, or one based on the exhaustion of the land, may be conducted by a person of very limited intelligence, nay, the most ignorant man may be fitted for the purpose, by the mere statement of facts to him. But a rational pursuit of agriculture, which, with the greatest economy of capital and labour, can obtain from a field continuously without exhaustion the highest crops it is capable of yielding, requires a large compass of knowledge, observation, and experience, more perhaps than in any other business. For the rational agriculturist must not merely know all the facts with which the illiterate peasant is acquainted, but he must also be able to appreciate them at their proper value; he must know the reason of all his proceedings, and what effect they may have upon his land. He must be able to interpret what his field tells him in the phenomena which he observes in practice; in a word, he must be a thorough man, and not a half-and-half creature who knows no more about his actions than a tom-cat, with just skill enough to catch gold fish in a basin of water.*

* If we compare the theoretical views expressed in the works of confessedly good practical farmers with the system of husbandry which they have found by their own experience to be the best, we observe the most irreconcilable contradictions between the two.

Walz ('Communications from Hohenheim,' No. 3, 1857) disputes both these propositions, viz.:—

'That the removal of the mineral constituents in the crops, *without compensation*, produces sooner or later lasting unfruitfulness as a consequence.'

'That if a soil is to maintain its fertility *continuously*, the removed mineral constituents must, sooner or later, be returned to it, i.e. the composition of the soil must be restored.'

And gives as his opinion that both these propositions are at present applicable only to soils of the worst kind, which needed a supply of mineral matters from the very beginning.

Now, if we turn to the 'Application of his theory to practice' (page 117), we would naturally suppose that he would never trouble himself about any compensation; but it soon appears that he is far from believing in the truth of his own doctrines. He lays the proper stress upon the restoration of potash, lime, magnesia, phosphoric acid, gypsum, guano, bone-earth, marl, and farm-yard manure; and lays down the following rule:—'That the farmer, to keep his ground in uniformly increasing fertility, must *not* remove *more* in his crops than the products of the atmosphere and the assimilable mineral substances added *annually* to the soil by the action of the weather.' He says further:—'If the farmer were to confine his business entirely, e.g. to the manufacture of *beer*, *spirit*, *sugar*, *starch-meal*, *dextrine*, *vinegar*, &c., and the sale of animal products merely to *butter*, using up the skimmed milk; if for his dairy he were to buy none but full-grown cows and not breed them himself, thus endeavouring to keep the phosphates upon his farm, then he would not only preserve continually the mineral substances in his store of manure, but he would also increase them by the yearly process of disintegration, unless he preferred to alienate the latter in his produce' (s. 142).

Hence the point of his practical teaching, in direct opposition to his theoretical, is that, in order to obtain uniform crops, great care must be taken to maintain and restore the composition of the soil.

The practical man proves that the notions which he has conceived are entirely inapplicable in his practice; and that the scientific principles which he disputes are precisely those by which he is unconsciously guided. Sound practice and true science are ever in unison; and a contest on these matters is possible only between two persons, one of whom does not understand the other. The chief fault lies in want of precision in defining things, and in using indefinite or vague language to express our ideas.

The opinion of Rosenberg-Lipinsky (see his 'Practical Agriculture,' b. ii. Breslau: E. Trewends, 1862), is 'that no kind of plant actually exhausts the great storehouse of the soil' (p. 738); and further, 'that plants, directly and indirectly, return to the soil more strength than they take from it' (p. 740). This opinion is thus modified (p. 742):—'when therefore the farmer does not take sufficient care that the more important magazine of nutriment, *the soil*, receives at the right time, and in proper quantity, the necessary compensation for that which is inevitably consumed, the picture of exhaustion which the cultivated plants manifestly wear, cannot possibly be charged upon their consumers, but the blame is *wholly* and *solely* attributable to the farmer himself.' Further, at p. 740, he says, 'Only in those plains, where the injustice of the elements, or of man, has violently disturbed the natural laws of the nutrition of plants, does the scanty vegetation of the wild flora indicate the exhaustion of the soil.'

CHAPTER XII.

COMMON SALT, NITRATE OF SODA, SALTS OF AMMONIA, GYPSUM, LIME.

Effect of these substances as elements of food ; their effect on the condition of the soil—Kuhlmann's experiments with common salt, nitrate of soda, and salts of ammonia ; experiments with the same substances in Bavaria ; conclusions : these matters are elements of food ; they are chemical means for preparing the soil ; they cause the distribution of the food in the soil in the form proper for the growth of plants—Experiments by Pincus with gypsum and sulphate of magnesia on clover; decrease of flowers and increase of stem and leaves of clover by sulphates ; the crop is not in proportion to the quantity of sulphates used—Effect of gypsum not yet explained ; indication in the comportment of clover soils with solution of gypsum ; such solution disperses potash and magnesia in the soil—Manures, their effect not explained by the composition of plants produced by them—Composition of the ash of clover manured with different substances—Effect of lime ; experiments of Kuhlmann and Träger ; comportment of lime-water with soils.

THESE salts are employed in agriculture in many cases with marked success as manure; and since nitric acid, soda, ammonia, sulphuric acid, and lime, are nutritive substances, the explanation of their efficacy presents no difficulty. But they also possess other peculiarities, by which they aid and promote the action of the plough and of mechanical tillage, as well as the influence of the atmosphere upon the condition of the field. This influence is not always clear to our minds, but it is not less certain.

We have every reason to believe that where the crops are increased by manuring with common salt alone, or when the favourable influence of salts of ammonia or nitrate of soda is augmented by the addition of common salt, the operation of the three salts essentially depends upon their power of diffusing the nutritive substances present in the soil, or of preparing those substances for absorption. In what manner this takes place with all is not yet explained. The first trustworthy experiments in this direction were made by F. Kühlmann (' Annal de Chim.' 3 ser. t. xx., p. 279). In the year 1845 he manured a natural meadow with sal ammoniac, sulphate of ammonia, and common salt; and obtained the following quantities of hay :—

Crop of hay per hectare, 1845 *and* 1846.

		Increased crop.
Unmanured	11263 kilos.	—
Sal ammoniac, yearly 200 kilos.	14964 "	3700 kilos.
" " 200 " / Common salt " 200 "	16950 "	5687 "

Another meadow yielded :—

Crop of hay, per hectare, 1846.

		Increased crop.
Unmanured	3323 kilos.	—
Sulphate of ammonia 200 kilos.	5856 "	2533 kilos.
" " 200 " / Common salt 133 "	6496 "	3173 "

For the purpose of examining the effect of common salt upon cereals, the General Committee of the Agricultural Society in Bavaria instituted at Bogenhausen and Weihenstephan, in the years 1857 and 1858, a series of experiments, conducted thus: of two plots, the one was manured with salts of ammonia, the other with the same quantity of salts of ammonia and an addition of 3080 grammes of common salt. These experiments were described at page 286, and it will be sufficient here to quote the crops which were obtained with salts of ammonia alone, and with common salt added to salts of ammonia.

Bogenhausen, 1857.

Barley.	Manured with salts of ammonia.		Manured with common salt and salts of ammonia.	
	Corn.	Straw.	Corn.	Straw.
	Grammes.	Grammes.	Grammes.	Grammes.
Plot I.	6355	16205	14550	27020
" II.	8470	16730	16510	36645
" III.	7280	17920	9887	24832
" IV.	6912	18287	11130	27969

Bogenhausen, 1858 (p. 287).

Winter-wheat.	Manured with salts of ammonia.		Manured with common salt and salts of ammonia.	
	Corn.	Straw.	Corn.	Straw.
	grammes.	grammes.	grammes.	grammes.
Plot I.	19600	41440	29904	61040
" II.	21520	38940	31696	71960
" III.	25040	57860	31416	74984
" IV.	27090	65100	34832	74684

In both these series of experiments, the crops of corn and straw were remarkably increased by the addition of common salt; and it is scarcely necessary to repeat, that such an augmentation could not possibly have taken place unless the soil had contained a certain quantity of phosphoric acid, silicic acid, potash, &c., capable of being brought into operation, but which without common salt was not assimilable.

Similar experiments were undertaken by the same society in Weihenstephan with nitrates; and the crops produced by these salts alone, and with the addition of common salt, per hectare, were as follows:—

Weihenstephan, 1857.—*Summer barley.*

	I. Unmanured.	II. Nitrate of soda.	III. Nitrate of soda with commonsalt.	IV. Nitrate of potash.	V. Nitrate of potash with commonsalt.	VI. Guano.
1857.	kilos.	kilos.	kilos.	kilos.	kilos.	kilos.
Summer-barley. Quantity of manure............	—	402	402+1379	473	473+1379	473
A { Corn	1604	2676	2366	2064	2313	1922
A { Straw........	2580	4378	4352	4219	4766	3300
1858. Winter-wheat. (the same manures.)						
B { Corn	1699	1804	2211	2248	2323	2366
B { Straw........	3030	3954	4151	4404	4454	5091

The experiments are remarkable in so far as they appear to indicate the cases in which the nitrates alone, or in combination with common salt, exert a favourable influence upon the increase of the crops.

The land in Weihenstephan is peculiarly suited for the cultivation of barley. Field A, after a manuring of the ordinary kind, about 600 cwt. per hectare, had borne turnips in 1854, peas in 1855, and wheat in 1856; it was then intended to let it lie fallow for one year, and to dress it at the end of the year for a new crop. On the other hand, Field B, before the experiment was made, had already borne four crops, namely, rape, wheat, clover grass, and oats; and was, in comparison with the first field, more exhausted, and by means of the oats and clover made much poorer in nutritive substances for the following cereal crop.

This seems to afford an explanation of the striking fact, that in 1857 the nitrates exercised upon the field a far more favourable influence than guano, although the soil had received as much nitrogen in the guano as in the nitrates, with the addition of phosphoric acid and potash. The field was still rich enough in nutritive substances for a good barley crop, and merely required their more uniform distribution (which was effected by the nitrates and the common salt), in order to make available to the roots of the barley plants as much or even more food than was the case with the plot manured with guano, on which the sum of the nutritive substances was greater.

In estimating the results of these experiments we must take into account the fact established by Dr. Zoeller, that soda seems to take a definite part in the production of barley seed. It is clear that the nitrates used did not simply act as agents in distributing other nutritive substances, but the soda as well as the nitric acid had their own share in the production of the crop. In the fourth experiment the field received as much nitric acid as in the second, but the base combined with the acid was potash and not soda; and in the fifth experiment the addition of common salt produced a remarkable increase in the corn crop. However, in the third and fifth experiments the quantity of salt applied was evidently too high, and the excess brought down the crop below that obtained with nitrate of soda alone.

Upon the more exhausted field in 1858 the crop obtained by guano in corn and especially in straw exceeded all the rest. In the arable soil of this field the amount of nutritive substances was on the whole smaller, and the addition of fresh elements of food made itself felt in a much higher degree than the distribution or dissemination of the substances already present in the soil. Still by the addition of common salt the crop of wheat was also increased.

The effect of potash upon wheat is as striking as that of soda upon barley.

As regards the effect of common salt and salts of soda generally, the analysis of the ash of turnips and potatoes, kitchen-garden and meadow plants, shows that, as a rule, the ashes of the former contain a considerable quantity of soda, and the ashes of the latter are proportionately rich in chlorides. The grass of a meadow, which has been manured with common salt, is eaten by cattle with greater relish, and preferred to any other, so that even from this point of view common salt deserves attention as a manure.

As that part of the action of nitrate of soda, sea-salt, and salts of ammonia, which consists in effecting the distribution in the soil of other elements of food, may consequently be replaced by careful tillage, the effect produced upon the crops by these salts affords a pretty safe indication of the condition of a field. If all other circumstances are the same, their effect will be much less marked upon a well tilled field than upon one not in the same condition.

Gypsum.—Among the recent investigations respecting the action of gypsum on clover,* those made by Dr.

* That excellent and most ably conducted agricultural journal, 'Zeitschrift des landwirthschaftlichen Vereins für Rhein. Preussen,' contains, in Nos. 9 and 10, September and October 1861, p. 352, the following statement about the remarkable fertility of a field for clover :—

'Twenty-three years ago Farmer Kirfield, of Rhon, in the hundred of

Pincus, of Insterburg, are the most important, both on account of the careful manner in which they were conducted, and the conclusions drawn from them. At Dr. Pincus' request, three plots of ground, each of a morgen (about $\frac{5}{8}$ of an acre) in extent, and lying close together, were selected by Mr. Rosenfeld in the beginning of May, from the middle of a large clover field in the neighbourhood of Lenkeningken. The clover crop had a very promising appearance, and the plants were then about an inch high. One of the plots was manured with a cwt. of gypsum, the second with the same quantity of sulphate of magnesia, and the intervening plant was left unmanured. The clover field from which the plots were selected was one of the best cultivated and most fertile in the district, and had produced in the preceding summer an abundant rye crop. The plants growing on the unmanured plot, when compared with those on the manured, very speedily presented a difference of colour and condition.

On the plot manured with gypsum, they were of a deeper green, and stood higher. The difference was most striking at the time of flowering, which occurred in the unmanured plots four or five days earlier than in the manured; the whole field being everywhere in full bloom, when scarcely a flower was to be seen in the manured plots. When the manured plots also were

Antweiler, Aldenau district (volcanic Eifel mountains), sowed a plot of land, said to abound in broken shells, with esparsette. For ten years he obtained good hay crops, and abundant after-grass. After this time a good deal of grass began to make its appearance among the esparsette. To destroy this Mr. Kirfield had his field deeply harrowed in spring, with iron harrows across the ridges, and then sown over again with 8 pounds of red clover seed. The red clover grew up splendidly with the esparsette, and gave for three years running two full crops per annum. At the end of the third year the land was again deeply harrowed and sown anew with 8 pounds of red clover seed. It gave again for three years running two full crops per annum of an excellent mixture of esparsette and red clover. The same operation was repeated twice after with the same success, so that the field has now for twenty-two years, consecutively, borne clover; that is to say, the first ten years esparsette alone, the following twelve years esparsette with red clover.'

It would be interesting to get a proper analysis of this soil, with especial regard to its absorptive power for potash and phosphate of lime.

in flower the clover was mown (May 24th). A square ruthe was measured from each of the experimental plots, and the clover separately cut and weighed.

Calculated per Prussian morgen (= $\frac{5}{8}$ of an acre), the results were,—

	Cwts. of clover-hay per morgen.
Without manure	21·6 cwts.
With gypsum	30·6 "
With sulphate of magnesia	32·4 "

On a closer examination of the clover-hay it was found that the increase in the crops obtained from the plots manured with the sulphates did not extend equally to all parts of the plant, but was more particularly observable in the production of stems. There were found in 100 parts of the clover from the manured plots more stems, fewer leaves, and still fewer flowers, than in 100 parts of the unmanured clover.

	Unmanured.	Manured with gypsum.	Manured with sulphate of magnesia.
100 parts of clover-hay, flowers	17·15	11·72	12·16
" leaves	27·45	26·22	25·28
" stems	55·40	61·62	63·0

or,

	Flowers.	Leaves.	Stems.
Clover-hay, unmanured	17·15	27·45	55·40
" manured with gypsum	11·72	25·28	63·0
" " sulphate of magnesia	12·16	26·22	61·62

These proportions of the different organs of the clover plant show that the action of the sulphates has led to a very considerable increase of the wood-cells, or, in other words, to an extension of the stems at the expense of the flowers and leaves. The relative proportion of the flowers, leaves, and stems, was :—

	Flowers.		Leaves.		Stems.
Clover-hay, unmanured	100	:	160	:	323
" manured with gypsum	100	:	216	:	507
" " sulphate of magnesia	100	:	216	:	538

According to the law of the symmetrical developement of plants, we may, without risk of error, take it for granted that the developement of the root increased in the same ratio as that of the stem. Now, as the increase of a plant in bulk is proportionate to the extent of food absorbent surface, we can understand that the manured plots should have produced when compared with the unmanured not only a larger mass of stems, but, as in the case of the sulphate of magnesia plot, also of flowers and leaves.

The entire crop per morgen, was,—

	Unmanured.	Manured with gypsum.	Manured with sulphate of magnesia.
Flowers	370·5 lbs.	358·5 lbs.	394·0 lbs.
Leaves	592·9 "	773·7 "	849·5 "
Stems	1196·6 "	1927·8 "	1996·5 "
	2160 "	3060 "	3240 "

The quantity of most of the ash constituents was found larger, nearly in the same proportion as the produce was greater. Phosphoric and sulphuric acids, however, showed in this respect a marked difference from the other ash-constituents, inasmuch as the quantity of these two substances was both absolutely and relatively larger in the clover from the manured plots.

The ash of the air-dried clover-hay amounted to—

	Unmanured.	Manured with gypsum.	Manured with sulphate of magnesia.
Per cent	6·95	7·96	7·94
In the entire crop	150·0 lbs.	243·0 lbs.	257·0 lbs.
Containing sulphuric acid	2·0 "	8·0 "	6·0 "
" phosphoric acid	11·95 "	21·55 "	21·82 "

The dressing with the sulphates had checked the developement of the flowers, and also that of the fruit; and it is evident that, though a higher crop of stems and leaves may be obtained by the use of these agents from a given surface, the result is not the same as regards the seeds. With an increase of flowers, leaves, and stems in the same ratio as on the unmanured plot, the two morgens of ground, dressed severally with gypsum and sulphate of magnesia, ought to have produced more than 600 lbs. of flowers each; whereas, compared with the enormous increase in the weight of the stems, and a not inconsiderable one in the weight of the leaves, we find no increase of flowers, and consequently also none of seed (Pincus). These most carefully conducted experiments confirm the general rule, that wherever external causes favour the developement of some organs, it can only be effected under like conditions of the soil, at the expense of other organs, and that in the case of clover, as in that of the cereals, increase of straw is attended with decrease of seed. (For further details of these experiments, see Appendix J.)

As the substitution of magnesia for lime, in the experiments now described, led to an increase of the clover crop, it may be safely assumed that in cases where gypsum is found to be favourable to the growth of clover, the cause must not be sought for in the lime, although it is very often found that many fields will grow clover only after a copious dressing with hydrate of lime. For we know also that gypsum promotes the growth of clover on many fields naturally abundant in lime; and since arable soil has the property of absorbing ammonia from the air and rain-water, and fixing it in the same or even a higher degree than salts of lime, there is only the sulphuric acid left to look to for an explanation of the favourable action of gypsum upon the growth of clover.

But the experiments of Pincus clearly demonstrate that the crops obtained by manuring with the sulphates bear no proportion whatever to the quantity of sulphuric acid supplied in them to the field.

The quantities of sulphuric acid severally contained in the two sulphates used were 30·12 lbs. in the sulphate of magnesia, and 44·18 lbs. in the sulphate of lime, which is as 6 : 8·8. The quantities of sulphuric acid in the two crops obtained severally by sulphate of lime and sulphate of magnesia, were as 6 : 8; the ash of the clover produced by sulphate of lime contained a little more than 8 lbs., and that from the sulphate of magnesia 6 lbs. On the plot dressed with gypsum the clover plant found a larger total quantity of sulphuric acid than on the sulphate of magnesia plot, and absorbed a correspondingly larger proportion. But this additional quantity of sulphuric acid absorbed did not increase the amount of produce; on the contrary, on the plot manured with sulphate of magnesia, which had received less sulphuric acid than the gypsum plot, the amount of vegetable matter was 8 per cent. higher than on the latter.

These facts show that we are still in the dark about the action of gypsum; and it will yet require a great many and most accurate observations before we are likely to arrive at a satisfactory explanation.

So long as the notion was generally entertained that plants derived their food from a solution, the effects of a soluble salt upon vegetation could, of course, be attributed only to the constituents of that salt. But now we are aware that the earth itself performs a special part in all the processes of nutrition; and there might, therefore, be grounds for supposing that the action of gypsum upon arable earth, or of the latter upon the former, might furnish a key, to some degree at least, to explain the effect of gypsum upon the growth of clover. A series of experiments made by me upon the alterations which a saturated solution of gypsum in water undergoes in contact with different arable soils, give very remarkable results, whch I will now state, without venturing to draw any definite conclusions from them.

I found that a solution of gypsum in contact with all the arable soils which I used, underwent decomposition, part of the lime separating from the sulphuric acid, and magnesia and potash taking its place, quite contrary to the ordinary affinities.

The experiments were made as follows:—300 grammes of each earth were mixed with a litre of pure water, and 300 other grammes of the same earth with a litre of a saturated solution of gypsum. After twenty-four hours the fluid was filtered, and the filtrate tested for magnesia. Pure distilled water took up from all the experimental earths, sulphuric acid and chlorine, besides traces of lime, magnesia, and soda, and occasionally also of potash, but mostly in inappreciable quantities. The alkalies, as well as the lime and the magnesia, seem to be dissolved by the agency of organic matters, as the dried residues blackened upon heating, and effervesced with acids after ignition.

Quantities of magnesia dissolved severally out of 300 grammes of earth by one litre of

	Distilled water. Milligr. of magnesia.	Gypsum water. Milligr. of magnesia.
Bogenhausen earth	30·2	70·6
Schleissheim "	31·6	87 8
Bogenhausen subsoil	12·2	84·2
Earth from Botanic Gardens	45·4	168·6
" Bogenhausen, No. I.*	26·6	101·6
" " No. II.	38·2	98·0
" Schornhof..............	8·6	63 4
" a cotton field, Alabama	1·9	3·8

These figures show that dressing a field with sulphate of lime makes the magnesia in the soil soluble and distributable. If the action which gypsum exercises upon the growth of clover depends really upon an increased supply of magnesia, this must surely be looked upon as one of the most curious facts known, since the increased supply is effected here by the aid of a lime salt. An experiment, made specially for the purpose, showed that the contact of arable earth with the solution of sulphate of lime is attended by an actual substitution of magnesia for lime; that is to say, a certain quality of lime is withdrawn from the solution and combines with the earth, whilst the liberated sulphuric acid, which was united to the lime, withdraws from the earth an equivalent quantity of magnesia. In a litre of gypsum water which had been in contact with 300 grammes of earth from a wheat-field, there were found the following quantities of sulphuric acid, magnesia, and lime:—

* On this field it had been experimentally proved that dressing with gypsum would give a larger clover crop. No. I. had not yet been manured with gypsum, No. II. had.

	The pure gypsum water contained in 1 litre—	The gypsum water which had been in contact with the earth—
Sulphuric acid............	1·170 grammes.	1·180 grammes.
Lime	0·820 "	0·736 "
Magnesia	— "	0·074 "

Besides the magnesia, a certain amount of potash also seems to be dissolved out of the earth by aid of the gypsum.

Out of 1000 grammes of earth from a wheat-field, there was dissolved by—

	3 litres of pure water.	3 litres of gypsum water.
Potash	24·3 milligr.	43·6 milligr.

These experiments show that the action of gypsum is very complex, and that it promotes the distribution of both magnesia and potash in the ground. This much is certain, that gypsum exercises a chemical action upon the soil, which extends to any depth of it, and that in consequence of the chemical and mechanical modification of the earth particles of certain nutritive elements become accessible to, and available for, the clover plant, which were not so before.

The cause of the action of a manuring agent is usually sought for in the composition of the plant, but I do not think that this is always to be relied upon. The composition of the seed of plants of wheat, for instance, is so constant, or varies so little, that it is quite impossible to infer from the results of the analysis of the seeds whether the soil on which they grow abounded or was deficient in phosphoric acid, nitrogen, potash, &c. The abundance or deficiency of food in a field exercises an influence upon the number and weight of the seeds, but not upon the relative proportion of their component elements. Thus, for instance, Pincus found a somewhat larger percentage of magnesia in the unmanured clover than in the plants manured with the sulphates; but taking the magnesia of the whole crop, the quantity of this substance was much larger in the latter than in the former.

Amount of magnesia in—

	Unmanured.	Manured with gypsum.	Manured with sulphate of magnesia.
100 parts of ash of clover-hay	5·87	5·47	5·27
In the whole crop	8·8 lbs.	13·29 lbs.	13·54 lbs.

Variations in the percentage proportions of potash, lime, and magnesia, may be often observed in all those plants in which, as in the case of tobacco, the vine, and the clover plant, potash may be substituted for lime, and vice versa. But in such cases the decrease of one body is invariably attended by a corresponding increase of the other.

Now if gypsum has the property of effecting a distribution of the potash in the ground, and this is wanting in magnesia, more potash should be contained in the clover manured with gypsum than with sulphate of magnesia. According to the analysis made by Pincus, the ash of the clover-hay contained :—

		Clover manured with gypsum.	Clover manured with sulphate of magnesia.
In per cent.	Potash..................	35·37 lbs.	32·91 lbs.
	Lime	19·17 "	20·66 "
In the whole ash	Potash..................	85·9 "	84·6 "
	Lime....................	46·6 '	53·2 "

These figures show that the quantity of potash is indeed larger, and that of lime smaller, in the crop produced by manuring with sulphate of lime than in the *higher* crop from sulphate of magnesia.

In the clover-hay reaped from the latter plot, the deficient potash was manifestly replaced by lime, and in the clover-hay from the gypsum manure plot, a certain amount of lime by potash.

An investigation, made with much carefulness, and without the least bias, as this by Pincus, appears, among the frivolous and loosely-conducted researches with which agriculture unfortunately abounds, like a green oasis in a dreary desert, and is well calculated to show how much real knowledge remains still to be gained of the processes in the soil with respect to the nutrition of plants. (See ' Agriculturo-chemical and

Chemical Researches and Experiments made by Dr. Pincus, at the Insterburg Station for Agriculturo-chemical and Physical Experiments.' Gumbinnen. 1861.)

Lime.—I have, unfortunately, never had an opportunity of examining a soil on which a lime-dressing has exercised a beneficial effect, as this substance is not used by farmers in the neighbourhood either of Giessen or of Munich. The experiments made by Kuhlmann, on meadows, in the years 1845 and 1846, seem to show that lime is principally useful in altering the condition of the soil; but having no data before me as to the particular soil on which these experiments were made, I am unable to point out wherein this alteration consists.

Hay crop reaped per hectare, 1845 *and* 1846.

	kilos.	kilos.
Meadow unmanured	11263	—
" manured with 300 kilos. of slaked lime, each year	14263	Increase 3000
" manured with 500 kilos. of chalk each year	10706	Decrease 556

It may safely be taken for granted here, that if the lime had acted as a nutritive element in the developement of the meadow plants, the plot manured with carbonate of lime ought to have given a higher, but assuredly in no case an inferior crop, than the unmanured plot. But the very reverse is the case: the carbonate of lime, which could only spread through the soil dissolved in carbonic acid, had an unfavourable effect; the caustic lime, on the contrary, was beneficial.

Among the Saxon experiments already so often alluded to, there are two of sufficient importance to deserve particular mention here. One of these was made by Traeger, of Oberbobritzsch; the other by Träger, of Friedersdorf. The latter omitted to make a comparative experiment to show the difference between the produce from a plot manured with lime, and that from an unmanured plot of the same size. Instead of the latter, therefore, I placed here by the side of the lime experiment, for the sake of comparison, another made with ground bones on a plot of the same size.

Experiment at Oberbobritzsch.

Lime manuring (110 cwts. quick lime).

	Produce per acre, unmanured.		Produce per acre, manured with lime.	
	Corn.	Straw.	Corn.	Straw.
	lbs.	lbs.	lbs.	lbs.
1851. Rye	1453	3015	1812	3773
1853. Oats	1528	1812	1748	2320
1852. Potatoes ..	9751	—	11021	—
1854. Clover-hay.	911	—	2942	—

Experiment at Friedersdorf.

Lime manuring (same quantity as above).

	Produce per acre, manured with 1644 lbs. ground bones.		Produce per acre, manured with lime.	
	Corn.	Straw.	Corn.	Straw.
	lbs.	lbs.	lbs.	lbs.
1851. Rye	990	3273	1012	3188
1853. Oats	1250	2226	1352	2280
1852. Potatoes ..	8994	—	12357	—
1854. Clover-hay.	4614	—	4438	—

Guano produced, in the year 1854, on the field at Oberbobritzsch, a higher clover crop than the lime (see page 266); but on the field at Friedersdorf it was smaller, 616 lbs. of guano produced, at Friedersdorf, 2337 lbs., at Oberbobritzsch, 5044 lbs., of clover-hay.

Experiments, in which I brought lime-water in contact with different samples of arable soil, have shown that the latter possesses a similar absorptive power for lime as for potash and ammonia. The earth was mixed with lime-water, and after remaining at rest until all alkaline reaction had disappeared, a fresh quantity of lime-water was then added, just sufficient to cause a feeble but permanent alkaline reaction.

Experiments on the amount of lime taken up out of lime-water by different arable soils.

	Lime out of lime-water.	
	grms. grains.	grms. grains.
1 litre* of Bogenhausen earth took up	2·824=43·5	2259=34788
1 " Schleissheim earth	2·397=37·0	1917=29521
1 " earth from Botanic Gardens	3·000=46·2	2400=36960
1 " subsoil from Bogenhausen .	3·288=50·6	2630=40502
1 " wheat soil	2·471=38·0	1976=30430
1 " from the same field after bearing a crop of clover	2·471=38·0	1976=30430
1 " of turf powder	6·301=97·0	5040=77616

The investigation into the alterations produced in the earth by the absorption of lime, more especially as regards potash and silicic acid rendered soluble, is not yet terminated.

* 1 Litre = 1 cubic decimètre = 61 cubic inches.

APPENDICES.

APPENDIX A (page 33).

EXAMINATION OF BEECH-LEAVES AT DIFFERENT STAGES OF GROWTH. (DR. ZOELLER.)

Beech leaves and asparagus, their ash-constituents at different periods of growth—The amylum of the palm—Motion of sap in plants—Drain, lysimeter, river, and bog water, their constituents—Fontinalis antipyretica from two different waters, ash-constituents—Vegetation of maize in solutions of its food—Experiments on the growth of beans in pure and prepared turf, results—Japanese agriculture—The cultivated soil of the torrid zone, its exhaustibility, its manure—Analysis of clover by Pincus—Clover sickness, its cause

THE beech tree (*fagus sylvatica*), from which the leaves examined were gathered, stands in the Botanical Garden of Munich. The leaves marked I. period were taken from the tree of four different sizes, on May 16, 1861. The smallest leaves *a* were just unfolded from the leaf-bud, whilst those marked *d* were fully expanded. There were between *a* and *d* a difference of four days' growth. The other two sets, marked severally *b* and *c*, were in size and period of growth intermediate between *a* and *d*. The leaves of the I. period were very delicate, and of yellowish green colour.

The leaves of the II. period were gathered on July 18, those of the III. period on October 15, 1861. The leaves of each period possessed among themselves the same size and firmness of structure. The colour of the July leaves was dark green, of those of October somewhat lighter.

The leaves of the IV. period were from the same tree, but had been gathered in the end of November, 1860. They had withered on the tree, and were quite dry.

One hundred parts by weight of the fresh beech leaves contained:—

	I. Period.				II. Period.	III. Period.
	a	*b*	*c*	*d*		
Dry substance	30·29	22·04	21·53	21·52	44·13	43·23
Water	69·71	77·96	78·47	78·46	55·87	56·77

One thousand fresh leaves contained, in grammes:—

Dry substance	10·01	15·90	32·63	60·00	116·16	117·53
Water	22·61	57·26	118·91	218·31	147·04	154·33
Total weight of 1000 leaves	32·62	73·16	151·54	278·31	263·20	271·86
Ash of dry leaves per cent.	4·65	5·40	5·82	5·76	7·57	10·15

The amount of water in the air-dried leaves of the IV. period was 11·89 per cent. The quantity of ash left by the dried leaves was 8·70 per cent.

For the ash analysis of the leaves of the period I., an equal number of leaves *b*, *c*, *d*, were incinerated.

One hundred parts of the ash of the leaves contained:—

	I. Period. 16th May, 1861.	II. Period. 18th July, 1861.	III. Period. 14th October, 1861.	IV. Period. End of Nov., 1860.
Soda	2·30	2·34	1·01	—*
Potash	29·95	10·72	4·85	0·99
Magnesia	3·10	3·52	2·79	7·13
Lime	9·83	26·46	34·05	34·13
Sesquioxide of iron	0·59	0·91	0·94	1·10
Phosphoric acid	24·21	5·18	3·48	1·95
Sulphuric acid	—*	—*	—*	4·98
Silicic acid	1·19	13·37	20·68	24·37
Carbonic acid and constituents not determined	28·83	37·50	32·20	25·35
Total	100·00	100·00	100·00	100·00

* Not determined.

Analysis of the ash of the leaves of the horse-chestnut and the walnut-tree, by E. STAFFEL. (*'An. der Chem. und Pharm.,'* vol. lxxvi. p. 372.)

	Horse-chestnut.		Walnut-tree.	
	Spring.	Autumn.	Spring.	Autumn.
Moisture in 100 parts of fresh substance, dried at 212° Fahr...	82·09	56·27	82·15	63·31
Per cents of ash in the fresh substance.....	1·376	3·288	1·092	2·570
Per cents of ash in the dried substance.....	7·69	7·52	7·719	7·005
100 *parts of ash contained—*				
Potash................	46·38	14·17	42·04	25·48
Lime..................	13·17	40·48	26·86	53·65
Magnesia..............	5·15	7·78	4·55	9·83
Alumina...............	0·41	0·51	0·18	0·06
Sesquioxide of iron.....	1·63	4·69	0·42	0·52
Sulphuric acid	2·45	1·69	2·58	2·65
Silicic acid............	1·76	13·91	1·21	2·02
Phosphoric acid........	24·40	8·22	21·12	4·04
Chloride of potassium...	4·65	8·55	1·04	1·73
Total..............	100·00	100·00	100·00	99·98

Analysis of the ash of flowering asparagus shoots, and of withered shoots with ripe fruit.—DR. ZOELLER.

	I. Flowering shoots.	II. Autumn shoots with ripe fruit.
Moisture in 100 parts of the fresh substance, dried at 212° Fahr........................	84·34	59·23
Per cents of ash of the fresh substance.........	0·946	4·13
Per cents of ash of the dried substance.........	6·050	10·13
100 *parts of ash contain—*		
Soda..	5·11	5·25
Potash..	34·40	11·77
Magnesia......................................	4·69	3·61
Lime..	9·07	24·05
Sesquioxide of iron............................	0·52	0·94
Phosphoric acid	12·54	7·33
Silicic acid....................................	1·85	9·68
Constituents not determined, &c..............	31·82	37·37
Total....................................	100·00	100·00

The asparagus shoots analysed came from the Botanical Garden at Munich. The flowering shoots were cut close to the ground, on June 20, 1861; the autumn shoots were cut in the same way, from the same plant, on October 28, 1861.

APPENDIX B (page 41)

ON THE STARCH IN THE STEMS OF PALMS.

The quantity of starch in one and the same stem differs to an extraordinary degree with the age of the plant, and the periods of flowering and fructification.

The generation of starch will in some instances rapidly increase not only within the cells, but occasionally even at the expense of the cellular tissue. Thus, in the root-stock of Sabal Mexicana, an abundance of starch is sometimes found, not only in the interior of the cells, but also outside the latter. But this phenomena is most striking in the East India Sago Palms (*Metroxylon*), in which it can be clearly observed that the generation of starch proceeds in distinct periods, and is in intimate organic connection with the development of the flowers and fruit. The Malays are in the habit of speaking of the tree as if it were with young at this period, during which it generates in its interior a large quantity of starch, forming the store of organic matter, out of which are to be produced, after liquefaction, new ligneous particles, and flowers, and fruit. This statement is peculiarly applicable to the *Metroxylon Rumphii Mart.* (*Sagus genuina Rumph*). This tree, which is a perfect chemical laboratory for the preparation of starch, is monocarpous, that is to say, it flowers and bears fruit only once, and then dies. It has by that time attained a height of from 28 to 30 feet. The stem, which is cylindrical, and more than a foot in diameter, consists of a mere shell, about one and a half to two inches thick, of a whitish wood of no great degree of hardness. Within the shell is enclosed a mass of spongy tissue formed of interlaced fibres, the cells of which are filled with starch granules. In the first stage of growth, whilst the stem still remains *unripe*, if the expression may be allowed, it contains only an inconsiderable quantity of starch. As growth progresses, and the base of the leaf stalks, and the upper part of the stem begins to be covered with long fibrous filaments or prickles, the quantity of starch increases.

The period of the greatest increase is indicated by the shedding of these prickles, and by the leaves being covered with a sort of white rime, as if powdered lime had been dusted over them. The Malays call this stage the *Maaputih*, i. e. the tree grows white. From the apex of the stem shoots forth at this stage the flower-stalk, which at a later period crowns the tree like an immense antler, bearing thousands of flowers, which are replaced afterwards by spherical fruit covered with scales. When the flower-stalk attains a length of one foot, the tree has entered that stage which the Malays term *Saga bonting*, that is with young. A small quantity of the starch is now taken up for the formation of the woody fibre of the flower-stalks. Finally arrives the period which the Malays term *Majang bara*, i. e. the young comes forth. The flower-stalk at the apex of the stem now attains a length of four feet, but the spathes out of which the floral branches are to project, are not yet opened. The tree may pass through these three stages without any great reduction of the store of starch; but at the next stage, termed *Batsja Bang*, i. e. the shoot branches out, when the flower stalk measures from six to ten feet in height, and ten feet in circumference, the greater portion of the starch is formed into thick woody fibre, and still more is this the case in the two last stages of the flower (*Siriboa*) and fruit (*Bahoa*), when there remains no longer any starch. A healthy tree produces between 400 and 800 lbs. of starch (the sago prepared from this is not sent to the European markets, but is consumed in the country). The palm, which produces the chief portion of the sago consumed in Europe, is the *Metroxylon laeve Mart.* of Malacca, the wild stems of which give four to five and a half picols of sago, whilst two to three picols only are obtained from those cultivated in gardens.

APPENDIX C (page 66).

VEGETABLE STATICS, LONDON, 1727.

The experiments made by Hales on the motion of the sap in vegetables, may be looked upon as the best model for all times of the most perfect method of investigation. That they are still at the present day unsurpassed in vegetable physiology may, perhaps, be attributed to the circumstance of their dating from the age of Newton. They deserve a place in every work treating of the physiology of plants.

In the beginning of his work Hales describes the experiments made by him on the motion of the sap in vegetables arising from the exhalation from their surface. These experiments were made with leafy branches, plants cut off from the roots, and others still retaining their roots.

The force of the pressure of a column of water, both with and without the cooperation of exhalation, was shown by the following experiment.

He fixed an apple-branch, three feet long, half-inch in diameter, full of leaves and lateral shoots, to a tube seven feet long, and five-eighths of an inch in diameter. He filled the tube with water, and then immersed the whole branch up to the lower end of the tube, in a vessel full of water. The water was driven into the branch by the pressure of the column of water in the tube, which subsided fourteen and a quarter inches in two days.

On the third day he removed the branch and tube out of the water, and hung it up in the open air; the water in the tube fell now twenty-seven inches in twelve hours.

To determine the comparative force with which the water is driven through the vessels of the ligneous body by pressure alone, and by pressure and exhalation combined, Hales joined a leafy apple branch to a tube nine feet long filled with water. In consequence of the pressure of the column of water and of the exhalation taking place from the surface of the leaves and twigs, the water in the tube (fortieth experiment) sank 36 inches in an hour. He then cut off the branch 13 inches below the glass tube, and placed the cut portion (with leaves and twigs) upright in a vessel with water. It was found to imbibe 18 ozs. of water in 30 hours; in which time only 6 ozs. of water had passed through the 13 inches of the stem connected with the tube, and that too under the pressure of a column of water 7 feet high.

In three other experiments, Hales shows that though the sap-vessels of plants will imbibe water plentifully by capillary attraction in branches severed from the trunk, as well as in those left in connection with the uninjured roots, they have very little power to protrude sap out at their extremities, and make it rise in a tube fixed to them.

The motion of the sap, Hales concludes, is to be attributed to the exhalation from the surface alone, and he proves that it proceeds in an equal degree from the trunk, branches, leaves, flower and fruit, and that the effect of the exhalation bears a certain definite ratio to the temperature and moisture of the air. When the atmosphere was charged with humidity little water was imbibed, and on rainy days the absorption was barely perceptible. Hales opens this second chapter of his statics with the following introductory remarks:—

'Having in the first chapter seen many proofs of the great quantity of liquid imbibed and perspired by vegetables, I propose in this to inquire with what force they do imbibe moisture.

'Though vegetables (which are inanimate) have not an engine which by its alternate dilatations and contractions does in animals forcibly drive the blood through the arteries and veins, yet has nature wonderfully contrived other means, most powerfully to raise and keep in motion the sap.'

In his twenty-first experiment he laid bare one of the chief roots of a thriving pear-tree at a depth of 2½ feet, cut off the end of the root, and connected the remaining stump with a glass tube filled with water and confined by mercury. This glass tube represents the root lengthened.

By the perspiration from the surface of the tree, the root imbibed the water in the tube with such vigor that in six minutes the mercury had risen in the tube as high as 8 inches, which corresponds to a column of water 9 feet in height.

This force is very nearly equal to that with which the blood moves in the great crural artery of a horse. 'I found,' says Hales, in his thirty-sixth experiment, 'the force of the blood of several animals, by tying them down alive upon their backs, then laying open the great crural artery where it first enters the thigh, and fixing to it, by means of two brass pipes running one into the other, a glass tube above ten feet long and one-eighth of an inch in diameter. In this tube the blood of one horse rose eight feet three inches, and the blood of another horse eight feet nine inches; the blood of a little dog, six feet and a half.'

Hales proved by special experiments, that the force of suction shown by him to be possessed by the roots of plants, is exercised equally by every individual branch, shoot, leaf, and fruit, in short, by every portion of the surface; that the motion of the sap from the root to the branches and leaves continues even when the trunk is, in any part, completely stripped of the outer and inner bark, and that this force of suction acts not only from the roots towards the top, but also from the latter towards the roots.

He concludes, from the results of his experiments, that every part of the plant is endowed with a powerful force of attraction.

We know now that it was not this force of attraction in itself that made the mercury and the water rise in Hales' tubes; and his experiments clearly show, that the imbibing force of plants, and of every leaf and root fibre, arising from surface exhalation, is aided by a powerful force from without, which is simply atmospheric pressure.

By the evaporation of the water from the surface of plants a vacuum is created therein, and in consequence thereof water and gases soluble in that fluid are readily forced in from without and raised by the pressure of the atmosphere, and it is this pressure from without which, together with capillary attraction, constitutes the principal cause of the motion and diffusion of the sap.

That the surface of plants possesses the faculty of imbibing gases, is most conclusively demonstrated by Hales. In his twenty-second experiment he says:—'The height to which the mercury rose in the tube did in some measure show the force with which the sap was imbibed, though not nearly the whole force; for while the water was imbibing, the transverse cut of the branch was covered with innumerable little hemispheres of air, and many air-bubbles issued out of the sap-vessels, which air did in part fill the tube as the water was drawn out of it; so that the height of the mercury could only be proportionable to the excess of the quantity of water drawn off, above the quantity of air which issued out of the wood.

'And if the quantity of air, which issued from the wood into the tube, had been equal to the quantity of water imbibed, then the mercury would not have risen at all, because there would have been no room for it in the tube.

'But if nine parts in twelve of the water be imbibed by the branch, and in the meantime but three such parts of air issue into the tube, then the mercury must needs rise near six inches, and so proportionately in different cases.'

When, in Hales' experiments, the root, the stem, or a branch had been wounded in any part by cutting off root fibres, or buds, or smaller twigs, the imbibing power was found to be diminished in the other parts (because at those wounded spots the difference in the pressure was more readily equalized by air finding its way in). The imbibing power was greatest about fresh cuts, but it gradually diminished until, after a few days, it remained no stronger about the cut than about the uninjured parts. Hales further concludes the exhalation from the surface to be the powerful cause that conveys nutriment to the plant from the parts surrounding it. If the proper proportion between the exhalation and the supply of food is in any way disturbed, the plant sickens and dies. If, in hot summers, the soil is unable to supply to the roots the moisture carried off in the course of the day by exhalation from the leaves, &c., and the tree or a branch of it is dried up, the motion of the sap ceases in such parts. Once dried up, the original action cannot be restored by capillary attraction alone. Exhalation is the chief condition of the life of the plant, serving as it does, to effect and maintain a continual motion of the sap, and a constantly recurring change in its condition.

'By comparing,' says Hales, 'the surface of the roots of a plant with the surface of the same plant above ground, we see the necessity of cutting off many branches from a transplanted tree. Suppose, upon digging the plant up, in order to transplant it, half the roots be cut off (which is the case of most young transplanted trees), then it is plain that but half the usual nourishment can be carried up through the roots, and that accordingly the perspiring surface above ground must be correspondingly reduced in order to restore the proper proportion between it and the imbibing surface under ground.' In the following observations on hop vines, Hales shows the effect of suppressed perspiration:—

'The soil of an acre of ground on which 9,000 hop-vines are growing, must supply to the plants, through the roots, in July, 36,000 ozs. of water in twelve hours. This is the quantity of water which during this time is exhaled by them, and which they must have to be in a thriving condition.

'In a kindly state of the air, this moisture is daily carried off in sufficient quantity to keep the hops in a healthy state; but in a rainy moist state of air, without a due mixture of dry weather, too much moisture hovers about the hops, so as to hinder, in a great measure, the kindly perspiration of the leaves, whereby the stagnating sap corrupts and breeds mould.

'This was the case in the year 1723, when ten or fourteen days almost continual rains fell, about the latter half of July, after four months' dry weather; upon which the most flourishing and promising hops were all infested with mould in their leaves and fruit, while the then poor and unpromising hops escaped and produced plenty; because they being small, did not perspire so great a quantity as the others; nor did they confine the perspired vapor so much as the large thriving vines did in their shady thickets.

'This rain on the then warm earth made the grass shoot out as fast as if it were in a hotbed; and the apples grew so precipitately, that they were of a very fleshy constitution, so as to rot more remarkably than had ever been remembered.

'The planters observe, that when mould has once seized any part of the ground, it soon runs over the whole, and that the grass and other herbs under the hops are infected with it; probably because the small seeds of this quick growing mould, which soon come to maturity, are blown over the whole ground; which spreading of the seed may be the reason why some grounds are infected with fen for several years successively.

'I have,' says Hales, 'in July (the season for fire-blasts, as the planters call them), seen the vines in the middle of a hop ground all scorched up, almost from one end of a large ground to the other, when a hot gleam of sunshine has come immediately after a shower of rain; at which time the vapors are often seen with the naked eye, but especially with reflecting telescopes, to ascend so plentifully as to make a clear and distinct object become immediately very dim and tremulous. Nor was there any dry gravelly bed in the ground, along the course of this scorch. It was, therefore, probably owing to the much greater quantity of scorching vapors in the middle than outside of the ground, and that being a denser medium, it was much hotter than a more rare medium.

'The gardeners about London have, to their cost, too often had occasion to observe a similar effect, when they have incautiously put bell-glasses over their cauliflowers early on a frosty morning, before the dew was evaporated off them; which dew being raised by the sun's warmth, and confined within the glass, did then form a dense transparent scalding vapor, which burnt and killed the plants.'

These observations translated into the language of the present day clearly show how acutely and exactly Hales comprehended the influence of perspiration upon the life of plants.

According to him, the proper thriving of plants depends upon the supply of food and moisture from the soil, which again is governed in a measure by a certain temperature and dryness of the atmosphere. The imbibing power of plants,—the motion of the sap in them, is dependent upon exhalation; the quantity of food imbibed and needed for the functions of the plant, is proportionate to the quantity of moisture exhaled in a given time. If the plant has imbibed a maximum of fluid, and the exhalation is hindered by a low temperature, or by long continued wet weather, the supply of food or the nutrition of the plant stops, the sap stagnates, and an alteration ensues tending to the generation of parasitical microscopic growths. If rain falls after hot weather, followed by a strong heat without wind, and every part of the plant is surrounded with an atmosphere saturated with moisture, cooling by further exhalation ceases, and the plants succumb to the sun-blasts.

APPENDIX D (page 98).

ANALYSES OF DRAINAGE, LYSIMETER, RIVER AND MARSH WATER.

I.—*Drainage Water.*

Thomas Way found in drainage water taken from seven different fields, the following constituents ('Journal of the Roy. Agric. Soc.,' vol. xvii. 133):—

	Grains in 1 gallon = 10,000 grains of water.						
	1	2	3	4	5	6	7
Potash	trace	trace	0·02	0·05	trace	0·22	trace
Soda	1·00	2·17	2·26	0·87	1·42	1·40	3·20
Lime	4·85	7·19	6·05	2·26	2·52	5·82	13·00
Magnesia	0·68	2·32	2·48	0·41	0·21	0·93	2·50
Sesquioxide of iron and alumina	0·40	0·05	0·10	—	1·30	0·35	0·50
Silicic acid	0·95	0·45	0·55	1·20	1·80	0·65	0·85
Chlorine	0·70	1·10	1·27	0·81	1·26	1·21	2·62
Sulphuric acid	1·65	5·15	4·40	1·71	1·29	3·12	9·51
Phosphoric acid	trace	0·12	trace	trace	0·08	0·06	0·12
Ammonia	0·018	0·018	0·018	0·012	0·018	0·018	0·006

Very similar results were obtained by Dr. Krocker in his analyses of drainage water from Proskau. (See Liebig and Kopp's 'Jahresbericht' for 1853, page 742.)

	Drainage Water (in 10,000 parts).					
	a	*b*	*c*	*d*	*e*	*f**
Organic matter	0·25	0·24	0·16	0·06	0·63	0·56
Carbonate of lime	0·84	0·84	1·27	0·79	0·71	0·84
Sulphate of lime	2·08	2·10	1·14	0·17	0·77	0·72
Nitrate of lime	0.02	0·02	0·01	0·02	0·02	0·02
Carbonate of magnesia	0·70	0·69	0·47	0·27	0·27	0·16
Carbonate of protoxide of iron	0·04	0·04	0·04	0·02	0·02	0·01
Potash	0·02	0·02	0·02	0·02	0·04	0·06
Soda	0·11	0·15	0·13	0·10	0·05	0·04
Chloride of sodium	0·08	0·08	0·07	0·03	0·01	0·01
Silica	0·07	0·07	0·06	0·05	0·06	0·05
Total solid matter	4·21	4·25	3·37	1·53	2·58	2·47

II.—*Lysimeter Water.*

Lysimeter water is atmospheric water passed by means of suitable apparatus (Lysimeter) through different soils, and collected after passing through. (See pp. 99, 100.)

The chemical analyses embraced four series, and were made by Dr. Zoeller.

1.—*Series of analyses made in* 1857.

The experiments were made with five different soils, 1 square foot of each earth, 6 inches deep, being placed in the several lysimeters. The quantities given represent the amount of atmospheric water that passed through the several lysimeters from April 7 to October 7, 1857. I. Manured calcareous soil, with vegetation (barley). II. Unmanured clay soil, with vegetation. III. Unmanured clay soil, without vegetation. IV. Manured clay soil, without vegetation. V. Manured clay soil, with vegetation. (2 lbs. cattle-dung, without straw, were severally used to manure the earth in lysimeters I., IV., and V.

* *a.* Drainage water from land A (a clay soil resting on a subsoil of calcareous loam or clay), collected 1st April, 1853.—*b.* The same, collected 1st May, 1853, after a heavy fall of rain (218 cubic inches on the square foot).—*c.* Drainage water from the same soil, mixed with drainage water from a humous clay soil, with calcareous clay or loam as subsoil, collected in October, 1853.—*d.* Drainage water from land B (tile-drained; subsoil of calcareous clay or loam), collected in October, 1853.—*e.* Water passing through the water-furrows from a heavy clay soil, collected in the beginning of June.—*f.* The same, collected in the middle of August, after heavy rains.

	I.	II.	III.	IV.	V.
	cub. cent.	cub. cent.	cub. cent.	cub. cent.	cub. cent.
Quantity of water passed through soil in lysimeter.	9845	18575	18148	19790	12302
	grammes.	grammes.	grammes.	grammes.	grammes.
Solid residue left at 212° F....	4·651	4·73	5·291	6·04	3·686
Ash of solid residue..........	3·127	3·283	3·545	4·245	2·610
Potash	0·064	0·044	0·037	0·108	0·047
Soda	0·070	0·104	0·135	0·470	0·074
Lime	1·436	1·070	1·285	1·354	1·136
Magnesia	0·203	0·165	0·024	0·058	0·063
Sesquioxide of iron..........	0·013	0·119	0·150	0·114	0·053
Chlorine	0·566	0·177	0·379	0·781	0·434
Phosphoric acid............	0·022	trace	trace	trace	trace
Sulphuric acid	0·172	0·504	0·515	0·580	0·412
Silicic acid...............	0·103	0·210	0·317	0·188	0·115
Clay and sand..............	0·089	0·074	0·112	0·045	0·047
Total..................	2·738	2·467	2·954	3·698	2·381
Deduct equivalent of oxygen corresponding to chlorine	0·127	0·040	0·085	0·176	0·095
Balance................	2·611	2·427	2·869	3·522	2·286
Carbonic acid and loss.......	2·040	2·303	2·422	2·518	1·400
Total..................	4·651	4·730	5·291	6·040	3·686

1,000,000 litres of water, passed through six inches of the soils already described, contain—

	I.	II.	III.	IV.	V.
	grammes.	grammes.	grammes.	grammes.	grammes.
Solid residue left at 212° F....	472·32	254·64	292·64	305·20	291·50
Ash contained in it..........	317·62	176·74	194·78	214·50	212·16
Potash....................	6·50	2·37	2·03	5·46	3·82
Soda	7·11	5·60	7·43	23·74	6·02
Lime......................	145·86	57·60	70·80	68·41	92·34
Magnesia..................	20·52	8·88	1·32	2·93	5·12
Oxide of iron..............	1·32	6·35	8·26	5·76	4·30
Chlorine..................	57·49	9·52	20·87	39·46	35·27
Phosphoric acid............	2·23	—	—	—	
Sulphuric acid	17·47	27·13	27·82	29·30	33·49
Silicic acid (soluble).........	10·46	11·35	17·46	9·50	9·34

2.—*Series of analyses made in* 1858.

The waters analysed were obtained from six soils, and represent the quantity of atmospheric water that passed, from May 10 to Nov. 1, 1858, through a layer of earth of a square foot of surface and 12 inches deep. The earth was ordinary unmanured alluvial lime soil from the Isar. The plant selected for cultivation was the potato. I. Unmanured, and without vegetation.

II. Unmanured, with vegetation. III. Manured, 10 grammes common salt, with vegetation. IV. Manured, 10 grammes nitrate of soda, with vegetation. V. 10 grammes guano, with vegetation. VI. Manured, 20 grammes phosphorite made soluble with hydrochloric (?) acid, with vegetation.

	I.	II.	III.	IV.	V.	VI.
	cu. cent.	cu. cent.	cu. cent.	cu. cent.	cu. cent.	cu. cent.
Quantity of water passed through the soil......	29185	25007	28138	17466	16520	30850
	grms.	grms.	grms.	grms.	grms.	grms.
Solid residue left at 212° F.	8·985	8·214	14·198	7·681	4·864	8·001
Ash of the solid residue...	6·591	6·094	12·292	5·553	3·704	6·192
Soda	0·250	0·245	3·290	1·255	0·301	0·233
Potash	0·075	0·066	0·034	0·035	0·032	0·029
Magnesia..................	0·432	0·443	0·454	0·264	0·382	0·374
Lime......	2·416	2·467	2·356	1·792	1·378	2·645
Oxide of iron............	0·115	0·083	0·104	0·083	0·096	0·117
Chlorine..................	0·227	0·237	3·925	0·177	0·317	0·238
Phosphoric acid..........	trace	trace	0·009	trace	0·007	0·015
Nitric acid..............	—	—	—	3·267	—	—
Sulphuric acid	0·132	0·147	0·118	0·182	0·197	0·666
Silicic acid..............	0·266	0·301	0·384	0·303	0·226	0·224
Sand	0·155	0·237	0·155	0·105	0·062	0·083
Sum................	4·068	4·226	10·829	7·463	2·998	4·644
Less the amount of oxygen equivalent to the chlorine	0·051	0·053	0·884	0·039	0·071	0·053
Sum................	4·017	4·163	9·945	7·424	2·927	4·591
Loss and carbonic acid....	4·968	4·051	4·253	0·257	1·937	3·410
Sum................	8·985	8·214	14·198	7·671	4·864	8·001

1,000,000 litres of water, passed through 10 inches of the soils already described, contain—

	I.	II.	III.	IV.	V.	VI.
	grms.	grms.	grms.	grms.	grms.	grms.
Solid residue left at 212° F.	307.86	328·46	504·58	439·76	294·42	259·35
Ash contained in it.......	225·83	243·69	436·84	374·04	224·21	200·71
Soda	8·56	9·79	116·92	71·85	18·22	7·55
Potash	2·56	2·63	1·20	2·00	1·93	0·94
Magnesia..................	14·80	17·71	16·13	15·11	23·18	12·12
Lime....................	82·78	98·65	83·73	102·59	83·41	85·73
Oxide of iron............	3·94	3·31	3·69	4·75	5·81	3·79
Chlorine..................	7·77	9·47	139·49	10·13	19·18	7·71
Phosphoric acid..........	—	—	0·31	—	0·42	0·48
Nitric acid..............	—	—	—	187·04	—	—
Sulphuric acid	4·52	5·87	4·19	10·42	11·09	21·59
Silicic acid..............	9·11	12·03	13·64	17·34	13·68	7·26

3.—*Series of analyses made in* 1859.

The waters analysed were obtained from six soils, and represent the quantity of atmospheric water that passed from March 20 to Nov. 16, 1859, through a layer of earth of a square foot of surface and 12 inches deep. The earth was ordinary unmanured alluvial lime soil from the Isar (garden soil). All the soils were in grass. I. Unmanured. II. Manured, 17·8 grammes nitrate of potash. III. Manured, 15·4 grammes sulphate of potash. IV. Manured, 17·8 grammes nitrate of potash, and 3·66 grammes phosphoride made soluble with 2 grammes sulphuric acid. V. Manured, 15·4 grammes sulphate of potash, and 3·66 grammes of phosphorite made soluble as above. VI. Manured, 12·3 grammes carbonate of potash.

	I.	II.	III.	IV.	V.	VI.
	cu. cent.	cu. cent.	cu. cent.	cu. cent.	cu. cent.	cu. cent.
Quantity of water passed through the soil......	20201	14487	20348	17491	23205	22488
	grms.	grms.	grms.	grms.	grms.	grms.
Solid residue left at 212° F.	4·5631	11·4272	15·1967	13·6805	20·784	5·5878
Ash of the solid residue...	3·192	8·861	13·644	10·681	17·668	4·614
Soda....................	0·044	0·069	0·083	0·030	0·085	0·038
Potash....................	0·024	0·166	0·205	0·231	0·244	0·112
Magnesia................	0·253	0·302	0·296	0·285	0·320	0·117
Lime....................	1·530	3·483	5·360	4·838	7·112	1·963
Oxide of iron............	0·072	0·057	0·072	0·084	0·088	0·053
Chlorine................	0·035	0·080	0·202	0·132	0·283	0·127
Phosphoric acid..........	trace	trace	trace	trace	trace	trace
Sulphuric acid...........	0·289	0·205	6·527	2·104	9·124	1·524
Nitric acid..............	1·125	5·913	1·301	5·248	1·401	1·390
Silicic acid..............	0·178	0·271	0·208	0·230	0·280	0·269
Sand....................	0·044	0·021	0·036	0·025	0·056	0·097
Sum................	3·594	10·567	14·290	13·207	18·993	4·690
Less the amount of oxygen equivalent to the chlorine............	0·007	0·018	0·045	0·029	0·063	0·028
Sum................	3·587	10·549	14·245	13·178	18·930	4·662
Loss and carbonic acid....	0·9761	0·8782	0·9517	0·5025	1·854	0·9258
Sum................	4·5631	11·4372	15·1967	13·6805	20·784	5·5878

1,000,000 litres of water, passed through one foot of the soils already described, contain—

	I.	II.	III.	IV.	V.	VI.
	grms.	grms.	grms.	grms.	grms.	grms.
Solid residue left at 212° F.	225·38	788·78	746·84	782·14	895·66	248·48
Ash contained in it.......	158·0	611·64	670·52	610·65	761·36	205·17
Soda....................	2·17	4·76	4·07	1·71	3·66	1·68
Potash..................	1·18	11·45	10·07	13·20	10·51	4·98
Magnesia................	12·52	20·84	14·54	16·29	13·79	5·20
Lime....................	75·73	240·42	263·41	276·59	306·48	87·29
Oxide of iron............	3·56	3·93	3·53	4·80	3·79	2·35
Chlorine.................	1·73	5·52	9·92	7·54	12·19	5·64
Sulphuric acid...........	14·30	14·15	320·76	120·29	393·19	23·30
Nitric acid..............	55·69	408·15	63·93	300·04	60·37	61·76
Silicic acid.............	8·81	18·70	10·32	13·14	12·06	11·96

4.—*Series of analyses made in* 1859, 1860.

This series is a direct continuation of the third. The waters analysed passed through the same soils through which the waters of the third series had already passed. The fourth series of experiments continued from Nov. 16, 1859, to April 12, 1860.

	I.	II.	III.	IV.	V.	VI.
	cu. cent.	cu. cent.	cu. cent.	cu. cent.	cu. cent.	cu. cent.
Quantity of water passed through the soil......	13500	12332	13760	13150	15232	14850
	grms.	grms.	grms.	grms.	grms.	grms.
Solid residue left at 212° F.	2·424	2·205	2·860	2·640	3·172	2·691
Ash of the solid residue...	2·071	1·682	2·395	2·086	2·599	2·220
Soda....................	0·021	0·024	0·028	0·022	0·028	0·019
Potash..................	trace	0·008	0·012	0·009	0·015	0·015
Magnesia................	0·065	0·058	0·069	0·074	0·070	0·063
Lime....................	0·770	0·859	1·016	0·938	0·952	1·057
Oxide of iron............	0·061	0·066	0·097	0·075	0·135	0·049
Chlorine.................	0·140	0·042	0·093	0·068	0·091	0·084
Phosphoric acid..........	trace	trace	trace	trace	trace	trace
Nitric acid..............	0·025	0·101	0·043	0·077	0·029	0·046
Sulphuric acid...........	0·119	0·099	0·487	0·474	0·527	0·185
Silica and sand*.........	0·170	0·144	0·118	0·153	0·123	0·136
Sum................	1·371	1·401	1·963	1·890	1·970	1·654
Deduct the amount of oxygen equivalent to the chlorine	0·024	0·009	0·020	0·015	0·020	0·018
Sum................	1·347	1·392	1·943	1·875	1·950	1·636
Loss and carbonic acid....	1·077	0·813	0·917	0·765	1·222	0·955
Sum................	2·424	2·205	2·860	2·640	3·172	2·691

* The quantity of sand very small.

1,000,000 litres of water, passed through 10 inches of the soils already described.

	I.	II.	III.	IV.	V.	VI.
	grms.	grms.	grms.	grms.	grms.	grms.
Solid residue left at 212° F.	179·56	178·80	207·71	200·81	208·24	181·21
Ash contained in it	153·47	136·39	174·07	158·69	170·62	149·49
Soda	1·56	1·94	2·04	1·73	1·83	1·27
Potash	—	0·64	0·92	0·69	0·98	1·01
Magnesia.................	4·86	4·70	5·02	5·56	4·59	4·24
Lime.....................	57·04	69·49	73·87	71·39	62·50	71·17
Oxide of iron.............	4·52	5·35	7·06	5·73	8·86	3·29
Chlorine....	10·43	3·40	6·76	5·21	5·97	5·65
Nitric acid...............	1·91	8·19	3·17	5·91	1·90	3·09
Sulphuric acid	8·86	8·02	35·45	36·08	34·59	12·45
Silicic acid with a little sand................	12·60	11·67	8·60	11·65	8·01	9·15

Compare 'Annal der Chem. und Phar.,' bd. 107, s. 27; 'Ergebnisse landwirthsch. und Versuche der Versuchstation, München,' II. Heft, s. 65, und III. Heft, s. 82.

Analysis of ashes of plants from the rivers Ohe and Iser.—Dr. Wittstein.

	Fontinalis from the Ohe.	Antipyretica* from the Iser.
Chloride of sodium	0·346	0·834
Potash	0·460	2·325
Soda....................................	1·745	
Lime	2·755	18·150
Magnesia	1·133	5·498
Alumina.................................	9·272	1·616
Oxide of iron............................	17·039	9·910
Oxide of manganese.......................	4·555	0·850
Sulphuric acid...........................	1·648	2·827
Phosphoric acid	trace	5·962
Silicic acid..............................	61·000	51·494
Carbonic acid	——	——
Sum....................................	99·953	99·466

* The great difference in the composition of the ashes of one and the same plant arises, according to Dr. Nägeli, less perhaps from the different amount of these matters in the water than from difference of age in the plants, and probably more still from other plants which nestle in the moss.

Analyses of river waters.

	WITTSTEIN.				H. S. JOHNSON.					
	From the Ohe.		From the Iser.		From the Regen.		From the Ilz.		From the Rachelsee	
	In 1000 grammes.	Per cent. of solid matter.	In 1000 grammes.	Per cent. of solid matter.	In 1000 grammes.	Per cent. of solid matter.	In 1000 grammes.	Per cent. of solid matter.	In 1000 grammes.	Per cent. of solid matter.
Chloride of sodium	0·00125	0·800	0·00163	0·723	0·0025	3·07	0·0059	6·52	0·0015	2·14
Chloride of potassium	0·00198	1·267	0·00413	1·832	0·0058[1]	7·13[1]	0·0043[1]	7·75[1]	0·0061[1]	8·73[1]
Potash	0·01282	8·205	0·00569	2·524	0·0096	11·80	0·0058	6·41	0·0123	17·59
Lime	0·00463	2·963	0·07830	34·737	0·0154	18·94	0·0092	10·17	0·0010	1·43
Magnesia	0·00165	1·056	0·01574	6·982	0·0026	3·19	0·0029	3·21	—	—
Alumina	0·00017	0·108	0·00030	0·133	0·0018[2]	2·21[2]	0·0052[2]	3·75[2]	0·0012[2]	1·72[2]
Oxide of iron	0·00037	0·237	0·02788	12·368	C·0009	1·10	0·0027	2·97	0·0012	1·72
Sulphuric acid	0·00182	1·165	0·00026	0·115	0·0020	2·46	—	—	—	—
Phosphoric acid	0·00525	3·360	0·00232	1·029	trace	trace	trace	trace	trace	trace
Silicic acid	0·01131	7·238	0·04955	21·981	0·0072	8·90	0·0095	10·50	0·0025	3·58
Organic matter	0·11500	73·601	0·03962	17·576	0·0335[3]	41·20[3]	0·0450[3]	49·72[3]	0·0441[3]	63·09[3]
Total quantity of solid residue	0·15625	100·000	0·22542	100·000	0·0813	100·000	0·0905	100·000	0·0699	100·000
Total quantity of inorganic matter	0·04125	—	0·18580	—	0·0478	—	0·0455	—	0·0258	—

[1] Soda. [2] Insoluble matter, sand. [3] Organic matter, carbonic acid. (Johnson, 'Annal. d. Chem. u. Pharm.' bd. xcv. s. 226.)

IV.—*Moss Water from the neighbourhood of Schleissheim.*—Dr. Wittstein.

The composition of the water was found to be as follows:—

	In 1000 grammes of water.	In 100 parts of solid matter.
Chloride of sodium	0·00280	1·101
Potash	0·00022	0·086
Soda	0·00551	2·167
Lime	0·05266	20·723
Magnesia	0·00921	3·627
Alumina	0·00029	0·114
Oxide of iron	0·00197	0·775
Sulphuric acid	0·00372	1·466
Phosphoric acid	0·00002	0·008
Silicic acid	0·00069	0·271
Carbonic acid	0·03943	15·595
Organic matter	0·13771	54·067
Total amount of solid matter	0·25423	100·000
Total amount of inorganic matter	0·11652	

APPENDIX E (page 113).

VEGETATION OF LAND PLANTS IN THE WATERY SOLUTIONS OF THEIR FOOD.

In experiments on the vegetation of land plants in the watery solutions of their food, great attention must be paid to the tendency of the fluid to become alkaline by the process of vegetation, as land plants always die in alkaline solutions. Great care must therefore be taken to keep the fluid neutral (very faintly alkaline) or feebly acid. Knop attained this object by frequently transferring his plants to fresh solutions; Stohmann, by placing the plants from the commencement in feebly acid solutions, and at a later period transferring them sometimes to fresh solutions, and at other times removing the alkaline reaction by frequent addition of a small quantity of acid.

The tendency of solutions to become alkaline by means of the plants vegetating in them, and the injurious effect of an alkaline solution on the growth of plants, were observed by Knop and Stohmann.

In the following are communicated the experiments of Knop and Stohmann on the vegetation of maize, in watery solutions.

I.—*Experiments of Knop.*

Knop based his experiments with maize on the earlier observations which he had made on the vegetation of barley and cresses (see 'Chem. Central Blatt,' 1861, s. 564). According to these observations the gramineæ require for their growth nothing more than a normal solution, which contains sulphate of magnesia, nitrate of lime, and nitrate of potash, according to the proportion $MgOSO_3 + 2CaONO_5 + 2KONO_5$, in which phosphate of iron was suspended, and phosphate of potash as required was dissolved. The normal solution A made according to the above formula contained in grammes—

	100 cent. cub.	500 cent. cub.	600 cent. cub.
Nitric acid	0·2160	1·0800	1·2960
Sulphuric acid	0·0495	0·2475	0·2970
Lime	0·0684	0·3420	0·4104
Magnesia	0·0233	0·1165	0·1398
Potash	0·0940	0·4700	0·5640
	0·4512	2·2560	2·7072

In consequence of using the solution in a more dilute form in the first period, in order to promote a better radication, 600 cubic centimètres of the above solution were employed at this time ; at every other period, 500 cubic centimètres were measured off, and to this last quantity the phosphate of potash was now added in the proportion indicated. The mixture, therefore, had the following composition in the five periods. The potash which was added as $KOPO_5$, and as $KONO_5$, are given separately and united with a bracket.

Period			
I.	12 cent. cub. solution of $KOPO_5$,*	600 cent. cub. normal solution A.	
II.	10 " "	500 " "	
III. & IV.	20 " "	500 " "	
V.	30 " "	500 " "	

In these solutions are contained in grammes,—

	Per. I.	Per. II.	Per. III & IV.	Per. V.
Nitric acid	1·2960	1·0800	1·0800	1·0800
Sulphuric acid	0·2970	0·2475	0·2475	0·2475
Phosphoric acid	0·0750	0·0625	0·1250	0·1875
Lime	0·4104	0·3420	0·3420	0·3420
Magnesia	0·1398	0·1165	0·1165	0·1165
Potash {	0·5640	0·4700	0·4700	0·4700
	0·0490	0·0408	0·0816	0·1224
	2·8312	2·3593	2·4626	2·5659

* 10 cent. cub. of the solution contained exactly 1 decigramme of $KOPO_5$

With the exception of the mixture used in Period V., there was added to the others also 0·1 gramme of phosphate of iron.

The duration of these periods was accidental, depending on fluctuating meteorological conditions of the atmosphere, but was so far regulated that a distant period was marked whenever almost exactly 1 litre of water had been exhaled through the leaves of the plants. At this time the remainder of the liquid was drawn off for analysis, and the vessel filled with a fresh solution.

In the following the results of the analysis are given along with the chief periods and circumstances of the experiments. In the analytical results in column A, is placed the total quantity of each acid, and salt received by the plant in that particular period; in column B, the bases and acids found by analysis in the remainder of the fluid; in column C, the difference between A and B, indicating the quantity of bases and acids absorbed by the plants. Further, the relations of the bases to each other, and that of magnesia to sulphuric acid (calculated from column A), are given; the quotients also express the proportions in which these matters were given to the plants at the beginning of the period. Immediately underneath, indicated by "absorbed," are placed the same proportions, calculated from column C, in order to show in what ratio the plant has selected these matters (when there does exist a determinate power of selection).

SUMMARY OF THE FOOD GIVEN TO A PLANT OF MAIZE AND ASSIMILATED BY IT

I. Period. From May 12 to June 12.—At the commencement the plant weighed 8 grammes*; and had six leaves with a surface of 264 square centimètres; water exhaled during the time = 1 litre. This period was divided into three sections, in which at first dilute solutions were used. The mixtures were in,—

	Section I.		Section II.		Section III.	
Solution of $KOPO_5$ . .	2	cent. cub.	4	cent. cub.	6	cent. cub.
Normal solution A....	100	"	200	"	300	"
Distilled water	198	"	96	"	—	"
Total fluid...........	300	"	300	"	306	"
Phosphate of iron	0·1	gramme.	0·1	gramme.	0·1	gramme

There were added as the solution was absorbed by the plant,—

* The maize seed were made to germinate in the month of April in well washed sand; the young plants weighed on the 12th May, 8 grammes; on drying the residue weighed scarcely more than the seeds.

I. Section	= 80	cent. cub. distilled water
II. "	= 350	" "
III. "	= 570	" "
	1000	cent. cub. = 1 litre

The residue from each section = 300 cent. cub., were united and analysed.

	A	B	C
Nitric acid	1·2960	?	?
Sulphuric acid	0·2970	0·1240	0·1730
Phosphoric acid	0·0750	0·0000	0·0750
Lime	0·4104	0·1480	0·2624
Magnesia	0·1398	0·0640	0·0758
Potash	0·6131	0·2280	0·3851
	2·8313	0·5640	0·9713

In the first of the following lines are placed the proportions of the matters given to the plants, calculated from column A; in the second, the calculations are made from column C:—

$$\text{Given: } \frac{CaO}{MgO} = 2{\cdot}9; \quad \frac{KO}{CaO} = 1{\cdot}5; \quad \frac{SO_3}{MgO} = 2{\cdot}1$$

$$\text{Absorbed: } \frac{CaO}{MgO} = 3{\cdot}4; \quad \frac{KO}{CaO} = 1{\cdot}5; \quad \frac{SO_3}{MgO} = 2{\cdot}2$$

II. Period. From June 12 to July 20.—At the commencement the plant weighed 65 grammes, and had nine leaves with a surface of 648 square centimètres; water exhaled = 1 litre; the plant received 0·1 gramme of phosphate of iron suspended in the water about the roots, the roots became of a reddish yellow colour.

	A	B	C
Nitric acid	1·0800	?	?
Sulphuric acid	0·2475	0·1704	0·0771
Phosphoric acid	0·0625	0·0000	0·0625
Lime	0·3420	0·1912	0·1508
Magnesia	0·1165	0·0860	0·0305
Potash	0·5110	0·3120	0·1990
	2·3595	0·7596	0·5199

Proportions of bases and acids,—

$$\text{Given: } \frac{CaO}{MgO} = 2{\cdot}9; \quad \frac{KO}{CaO} = 1{\cdot}5; \quad \frac{SO_3}{MgO} = 2{\cdot}1$$

$$\text{Absorbed: } \frac{CaO}{MgO} = 5{\cdot}0; \quad \frac{KO}{CaO} = 1{\cdot}3; \quad \frac{SO_3}{MgO} = 2{\cdot}5$$

III. Period. From July 20 to 27.—At the commencement the plant weighed 73 grammes, and had eleven leaves with a surface of 720 square centimètres; water exhaled = 1 litre; to the solution was added 0·1 gramme of phosphate of iron; radication strong. This period differs from the preceding in the quantity of $KOPO_5$ given being double.

	A	B	C
Nitric acid	1·0800	?	?
Sulphuric acid	0·2475	0·1716	0·0759
Phosphoric acid	0·1250	0·0000	0·1250
Lime	0·3420	0·1440	0·1980
Magnesia	0·1165	0·0860	0·0305
Potash	0·5518	0·2160	0·3358
	2·4628	0·6176	0·7652

Proportions of bases and acids,—

$$\text{Given: } \frac{CaO}{MgO} = 2{\cdot}9; \quad \frac{KO}{CaO} = 1{\cdot}5; \quad \frac{SO_3}{MgO} = 2{\cdot}1$$

$$\text{Absorbed: } \frac{CaO}{MgO} = 6{\cdot}1; \quad \frac{KO}{CaO} = 1{\cdot}7; \quad \frac{SO_3}{MgO} = 2{\cdot}4$$

IV. Period. From July 27 to August 1.—At the commencement the plant weighed 147 grammes, had eleven leaves, with a surface of 1160 square centimètres; water exhaled = 1 litre; to the solution was added 0·1 gramme of phosphate of iron; the roots became distinctly reddish yellow. The plants received twice as much $KOPO_5$ as in the second period.

	A	B	C
Nitric acid	1·0800	?	?
Sulphuric acid	0·2475	0·1374	0·1101
Phosphoric acid	0·1250	0·0000	0·1250
Lime	0·3420	0·1188	0·2232
Magnesia	0·1165	0·0719	0·0446
Potash	0·5518	0·1296	0·4222
	2·4628	0·4617	0·9211

Proportions between bases and acids,—

$$\text{Given: } \frac{CaO}{MgO} = 2{\cdot}9; \quad \frac{KO}{CaO} = 1{\cdot}6; \quad \frac{SO_3}{MgO} = 2{\cdot}1$$

$$\text{Absorbed: } \frac{CaO}{MgO} = 5{\cdot}0; \quad \frac{KO}{CaO} = 1{\cdot}8; \quad \frac{SO_3}{MgO} = 2{\cdot}3$$

To ascertain how far the results from this artificial mode of cultivation may be compared with those produced under natural circumstances, maize of the same kind was planted in the garden in the middle of May. The latter were exposed to the same atmospheric conditions as the experimental plants. On August 1, a plant from the garden of the same period of vegetation as the experimental plant, with also fifteen leaves, and visible male flowers, weighed 1260 grammes, that is to say, seven times as much as the artificially reared plant. The stem of the garden plant from the lower knot to the summit of the flower-stalk measured 150 centimètres, being three times the height of the experimental plant.

V. Period. From August 1 to 10.—At the commencement the plant weighed 173 grammes; the stem was 52 centimètres high; in the middle of the period the plant had fifteen large fine green leaves, with a surface of 1420 square centimètres. In this period double the quantity of water (2 litres) was exhaled, and as the older roots were distinctly reddish yellow in colour, the plant received no more phosphate of iron, but thrice as much phosphate of potash as in the second period. On August 6 and 7, the male flower, consisting of seven single ears, was fully expanded from the sheath, the stem was strong, and 70 centimètres high. On August 7, a fully formed female flower appeared; on August 9, the anthers began to shed their pollen.

	A	B	C
Nitric acid	1·0800	?	?
Sulphuric acid	0·2475	0·1640	0·0835
Phosphoric acid	0·1875	0·0020	0·1855
Lime	0·3420	0·1236	0·2184
Magnesia	0·1165	0·0790	0·0370
Potash	0·5927	0·1894	0·4033
	2·5662	0·5580	0·9277

Proportions between bases and acids,

$$\text{Given: } \frac{CaO}{MgO} = 2·9; \quad \frac{KO}{CaO} = 1·7; \quad \frac{SO_3}{MgO} = 2·1$$

$$\text{Absorbed: } \frac{CaO}{MgO} = 5·9; \quad \frac{KO}{CaO} = 1·8; \quad \frac{SO_3}{MgO} = 2·3$$

As the plant in this period flowered, and earlier experiments had shown that maize dug up at the period of flowering, and placed in river water furnished still ripe seeds, and also by the addition of the salts which the plant in each period had taken up in proportion to its increase in weight in the first four periods, it appeared that it must contain fully as much salts as the plant in its normal condition in the field takes up, if placed from this period only in distilled water.

VI. Period. From August 10 to 16.—At the commencement the plant weighed 255 grammes, and had 15 fully expanded leaves with a surface of 2640 square centimètres: 2 litres of water were exhaled.

On August 10, the anthers had almost completely shed their pollen. The stem shot up rapidly, and on the 12th it measured to the tip of the flower 1 mètre in height. On the 13th a second female flower appeared, which was surrounded with paper to protect it from dust. On August 16 the height of the plant was 1·1 mètre; it did not grow any more. The fruit-bearing stalk was, on August 16, already 2 decimètres long, and had below a thickness of 4 centimètres.

On August 16 the water was drawn off and analysed.

Present.	Not present.
0·016 gramme potash.	Sulphuric acid (only indistinct opalescence with chloride of barium).
0·008 " lime.	Magnesia.
0·001 " phosphoric acid.	Iron and silicic acid.

From the circumstance that in this solution there was no silicic acid, it is plain that the glass vessel had furnished none to the fluid by decomposition in the course of one to two weeks.

VII. Period. From August 16 to September 4,—

Weight of plant on	16 August	....	280 grammes.
" "	22 "	at 9 o'clock a.m.	316 "
" "	22 "	" 9 " p.m.	320 "
" "	28 "	" 9 " "	330 "
" "	1 Sept.	" 9 " "	327 "
" "	4 "	" 9 " "	317 "

From September 1 the weight diminished by the drying of the leaves, and as this decrease was accidental, the plant was not thenceforward weighed. The leaves shrivelled. The plant had exhaled 3½ litres of water in the period. At this time it was placed in a vessel containing 1·5 litres of water, to determine what salts returned to the water by endosmone. The water was kept up at the same level by daily additions, and at last was allowed to exhale until the residue was 1 litre. In this litre were found 0.031 carbonate of lime, and 0.007 carbonate of magnesia. Both salts were left in the basin undissolved after evaporation, and after the residue had been treated with water.

In the water with which the residue left on evaporation in the basin had been extracted, the following substances were found in solution:—

0·020 lime	together with organic matter which reduced a solution of oxide of copper and potash.
0·0006 phosphoric acid	
0·0034 potash	

In this last solution not a trace of iron, sulphuric acid or magnesia was found. As the preceding analyses indicate, the solution of nutritive matters for gramineæ must have the following composition:—

$$MgOSO_3 + 4CaONO_5 + 4KONO_5 + xKOPO_5$$

(Compare 'Chem. Central Blatt, 1861,' s. 465, 564, and 945.)

* At all periods the plants threw off organic substances, but chiefly in the last periods.

II.—*Experiments of Stohmann.*

The experiments of Stohmann agree in their main results with those of Knop. According to these experiments, the maize plant grows to full maturity if in the beginning of May the seed which has germinated in water, and has shot forth roots, is placed in a solution containing the food of maize in the proportions in which they exist in the ashes, if at the same time there has been added to it so much nitrate of ammonia that to every part of phosphoric acid in the solution there are two parts of nitrogen, and if finally it has been diluted with distilled water to a concentration of three parts of solid matter per 1000 parts. The plants must grow in a sunny spot, and the water exhaled by the leaves must be daily replaced by distilled water, and the solution tested as to its reaction. The solution must always react, slightly acid, and be maintained in this condition by the addition from time to time of a few drops of phosphoric acid. If these conditions are fulfilled, there is no necessity for any artificial source of carbonic acid, but by means of the atmospheric carbonic acid alone there are produced fully formed plants which, under favourable circumstances, attain a height of 7 feet.*

The experiments of Stohmann were more especially directed to the influence exercised on the growth of the maize plant by the withdrawal of one element of food. In this point the results differ from those of Knop. Whilst in the experiments of the latter, maize was found to grow perfectly without silicic acid, soda, or ammonia, Stohmann made use of silicic acid in all his experiments, and found further that by the complete withdrawal of ammonia and even soda the plants grew quite well.

On withdrawing *ammonia* completely and replacing it by nitric acid, Stohmann found that the plants grew perfectly well for the first ten to twelve days, then they became of a pale yellowish green, and the vegetation proceeded extremely slowly.

If after a month's vegetation a little ammonia (in the form of nitrate or acetate) was given to the plants, they died very quickly. Without this supply of ammonia the blanched, sickly vegetation continued; the plant did not die, and yet it could not be said to live.† In the experiments made without *soda*, it was found that the plant could dispense with this substance at first, but its progress was soon arrested if the soda was completely withdrawn. The nitrate of lime of the normal solution was in another experiment replaced by a corresponding quantity of nitrate of magnesia. The growth of the maize plant was after a short time much retarded, only a few small, thin leaves being developed. By the addition of a little nitrate of lime to the growing plant, the most

* According to Knop maize plants growing in a watery solution give off carbonic acid continuously from their roots.

† Compare Knop, 'Chem. Central Bl. 1862,' s. 257.

remarkable change was however produced. Scarcely five hours elapsed before the growth of the plant, which had been stationary for four weeks, awakened to a new life, and proceeded from this time forth in the best manner possible. A plant without the after addition of nitrate of lime remained stationary, making no progress whatever: the maize plant, therefore, requires lime immediately after the commencement of its growth.

In an experiment in which the *magnesia* was replaced by nitrate of lime, the same result was obtained as when lime was wanting. In this case, also, the vegetation was very poor. A supply of magnesia in the form of nitrate, exerted here also the most favourable action, only the effect was not so quickly produced as in the case of lime.

Even by the complete withdrawal of *nitric acid* the maize plant did not grow. In these experiments it is true the alkalies, as well as the alkaline earths, were in part supplied in the form of sulphates and chlorides. Chlorine and sulphuric acid, however, are required only to a limited extent in the vegetable organism. The same holds good in the experiment without nitrogen. According to these experiments, therefore, a plant is not developed if one of its elements of food is wanting, and the complete replacement of one element of food by another one similar to it, is hence completely out of the question. The result may, however, be different with the reciprocal *partial* replacement of similar elements of food; and Stohmann is about to take up this question.

The form in which the food was supplied was the following.* The *silicic acid* was always supplied in the form of silicate of potash; the potash as nitrate. In the series of experiments (3) which were made without nitric acid, sulphate of potash was used instead of the nitrate.

The *phosphoric acid* was used in the form of $2NaO, HO, PO_5 + 24HO$; in experimental series 5, in which soda was excluded, a potash salt was used, $2KO, HO, PO_5$, of which a concentrated solution was prepared, containing a known quantity of potash and phosphoric acid. As the phosphate of soda contained more soda than was requisite in the composition of the ash, there was thus in the fluids in the experimental series 1 to 7 an excess of this base; at a later period, a correspondingly smaller quantity of phosphate of soda and more of the potash salts were employed.

The *sulphuric acid* was in the form of sulphate of magnesia, with the exception of 7, in which sulphate of ammonia was used, the magnesia required was added in the form of nitrate of magnesia.

The *oxide of iron* was supplied in the form of pure sublimed-

* To form a complete solution of all matters, and to remove the alkaline reaction, the fluid was first properly diluted with water and so much weak hydrochloric and later phosphoric acid was added as to make the reaction distinctly feebly acid.

chloride; the *lime* as nitrate, and in the case of 3 as chloride of calcium; the *ammonia* as nitrate, sulphate, or chloride.

It was scarcely possible to avoid using a larger or smaller excess of one or other of the substances. This was particularly the case with soda and chlorine. These deviations will be best shown in the following tables.

Experimental series.

	Intended composition.	1 Normal.	2 Without ammonia.	3 Without nitric acid.	4 Without nitrogen.	5 Without soda.	6 Without lime.	7 Without magnesia.
Potash	35·9	35·9	52·0	35·9	35·9	35·9	35·9	35·9
Soda	1·0	8·0	8·0	8·0	8·0	—	1·0	1·0
Lime	10·8	10·8	10·8	10·8	10·8	10·8	—	19·2
Magnesia	6·0	6·0	6·0	6·0	6·0	6·0	13·7	—
Oxide of iron	2·3	2·3	2·3	2·3	2·3	2·3	2·3	2·3
Sulphuric acid	5·2	5·2	5·2	26·9	26·9	5·2	5·2	5·2
Chlorine	1·3	19·7	3·1	66·5	16·8	3·1	3·1	3·1
Phosphoric acid	9·1	9·1	9·1	9·1	9·1	9·1	9·1	9·1
Silicic acid	28·5	28·5	28·5	28·5	28·5	28·5	28·5	28·5
Nitrogen	18·2	18·2	18·2	—	18·2	18·2	18·2	18·2

Summary of the weights of the crops.

Experimental series.	Plants.	Parts of plants.	Dry substances.	Amount of ash.	Amount of ash.	Organic matter.	Proportion of the weight of the seed to that of the crop after deduction of the ash.
			grms.	grms.	per ct.	grms.	
	From the garden.	Roots	10·36	15·24	11·4	—	—
		Stem	52·39				
		Leaves	42·39				
		" of the head.	28·51				
		Grains	190·14	3·42	1·8	—	—
		3 heads	22·66	0·54	2·4	—	—
		Entire plant	346·45	19·20	5·5	327·25	1 : 3147
1.	A	Roots	3·92	3·97	13·1	—	—
		Stem	9·67				
		Leaves	11·79				
		" of heads	4·91				
		Head with grain	34·09	0·82	2·4	—	—
		Entire plant	64·38	4·79	7·5	59·59	1 : 573
	B	Straw	27·36	4·35	15·9	—	—
		Heads	4·24	0·14	3·4	—	—
		Grains	24·57	0·56	2·3	—	—
		Entire plant	56·17	5·05	8·9	51·12	1 : 491

Summary of the weights of the crops.—(Continued.)

Experimental series.	Plants.	Parts of plants.	Dry substances.	Amount of ash.	Amount of ash	Organic matter.	Proportion of the weight of the seed to that of the crop after deduction of the ash.
			grms.	grms.	per ct.	grms.	
	C	Entire plant	55·52	5·94	10·7	49·58	1 : 477
2.	D	" "	62·44	6·49	10·4	55·95	1 : 538
	A—C	" "	1·19	—	—	—	—
	D	" "	2·39	0·54	22·8	1·85	1 : 18
3.	A—B	" "	0·204	—	—	—	—
4.	A	Roots	0·45	0·10	22·8	—	—
		Stem and leaves...	1·03	0·17	16·7	—	—
		Entire plant	1·48	0·27	18·2	1·21	1 : 12
	C	" "	10·90	0·92	8·5	9·98	1 : 96
	D	" "	39·48	5·57	14·1	33·91	1 : 326
5.	A	" "	49·63	5·21	10·5	44·42	1 : 427
	B	" "	32·31	3·36	10·4	28·95	1 : 278
6.	A	" "	0·30	—	—	—	—
	B	" "	84·30	8·22	9·75	76·08	1 : 731
7.	A	" "	0·82	0·18	21·4	0·64	1 : 6
	B. C	" "	6·01	0·82	13·7	5·19	1 : 50

REMARKS ON THE SUMMARY OF THE WEIGHTS OF THE CROPS.

I. Plants A, B, C, and D grew in normal solutions. Plants A and B were placed in the solution on July 1, and plant A was gathered on September 10, fully ripened; its total height was 202 centimètres. The plant from the garden soil with which it was compared was of middle size. Plant B gathered on September 27, was fully grown, and had a height of 127 centimètres. Plants C and D were placed in the normal solution on June 10, they did not attain their full growth; both were gathered on October 28.

II. Commencement of experiment in solutions *without ammonia* on June 10.—A and B received on July 12 a supply of 0·2 gramme nitrate of ammonia; on July 23 they were placed in a fresh solution, to which was added 0·2 gramme acetate of ammonia; both plants died on July 31. Plants C and D received normal solution on August 4, which was neutralised with phosphoric acid; C died on August 9, D recovered somewhat, but remained sickly till gathered on September 27.

III. Experiments *without nitric acid.*—Commencement on June 10; rapid decay of the plants; by July 1 A and B were already dead.

IV. Experiments *without nitrogen.*—Commencement on June 10. In the first week the growth was excellent, but in the second week it came to a stand. A lived till gathered on September 27; height 15 centimètres, length of roots 82 centimètres. Plants C and D received on July 11 each 0·2 gramme nitrate of ammonia, and on July 17, also, the same quantity. The influence of this salt was rapidly visible. On August 4, C and D received normal solution. Plant C was gathered on September 27, height 75 centimètres. Plant D, gathered on November 15, was in a healthy state, and had attained a height of 120 centimètres.

V. Experiments *without soda.*—Commencement June 10. The early vegetation was very luxuriant; in the end of July, however, the plants were not progressing. On August 4, the plants received normal solution; two died, but A and B made further progress. A and B were gathered on October 30, height of A, 205 centimètres; B stunted.

VI. Experiments *without lime.*—Commencement June 10. Plant A had reached a height of 2 centimètres on July 17; but made no further progress. B received on July 1, 0·1 gramme lime in the form of nitrate, and on August 4, normal solution, vigorous growth. It had on November 15 four stems respectively 107, 95, 75, 70 centimètres high, which were covered with leaves, and had eight well developed heads of fruit.

VII. Experiments *without magnesia.*—Commencement June 10. Progress as in Experiment VI., and gathered as it was making no visible progress. B and C received on July 17, 0·1 gramme magnesia, and on August 4 normal solution, gathered September 27; height of B, 23 centimètres; of C, 42 centimètres. Both had male flowers without pollen, and no female flowers.

On comparing his experimental plants with those which grew in the ground, both in respect to weight of the crop and to amount of ash and its composition, Stohmann concluded that we may indeed convert a plant of maize into a *water-plant*, but that maize cannot grow in a normal condition in solutions of its food. Further, his experiments showed in a positive manner that the soil played a determinate part in the nutriment of plants—absorption of alkalies—and that plants in the absorption of their food must themselves take an active part (compare Henneberg's 'Journal für landwirthschaft, 1862,' s. 1. and 'An. der Chem. und Pharm.,' bd. cxxi. s. 285).

APPENDIX F (pp. 114, 115).

EXPERIMENTS ON THE GROWTH OF BEANS IN POWDERED TURF.

To complete the experiments on vegetation described at page 112, the results of the entire crops are now given in the following table:—

Dry substance of the bean plants in grammes.

	I. Pot, fully saturated.	II. Pot, half saturated.	III. Pot, quarter saturated.	IV. Pot, pure turf.
Seed	93·240	66·127	50·463	7·069
Shell	25·948	18·393	13·658	2·631
Leaves	19·420	15·797	12·477	1·979
Stem	26·007	20·107	15·710	5·676
Roots	58·399	36·368	25·411	3·063
Total weight	223·014	156·792	117·719	20·418

These numbers completely confirm the conclusions drawn from the weight of the seeds alone. If the crop from the pure turf be taken as unity, the weights of the entire crops bear the following proportions—

1 : 5·7 : 7·7 : 10·9

or if the weight of the crop in the $\frac{1}{4}$ saturated turf be called 2, and that of the $\frac{1}{2}$ and fully saturated turf be compared with it, the following proportions are found—

2 : 2·7 : 3·8

If the weight of the crop furnished by the pure turf be subtracted from each of the others, and the weight of the crop in the $\frac{1}{4}$ saturated turf be taken at 2, then the crops in the $\frac{1}{2}$ and fully saturated turfs bear the following proportions to it—

2 : 2·8 : 4·2

APPENDIX G (page 229).

Extract from the Report to the Minister of Agriculture at Berlin, on Japanese Husbandry; by Dr. H. Maron, *Member of the Prussian East Asiatic Expedition.*

SECTION I.

SOIL AND MANURING.

The Japanese empire stretches from the 30th to the 45th degree of north latitude. The average temperature and distribution of heat constitute a climate embracing all the gradations between those of central Germany and of Upper Italy. A solitary tropical palm, not fully developed, grows by the side of the northern pine, rice and cotton along with buckwheat and barley. Everywhere on the chains of hills, which cover the whole country like an irregular fine network, the pine predominates, stamping upon the landscape that homely northern character, which affords so cheering a sight to the northern traveller, who reaches these shores after having passed through the hot and luxuriant regions of the tropics. In the valleys, on the other hand, the burning south holds sway, covering the ground with a rich vegetation of rice, cotton, yams, and sweet potatoes. Hundreds of footpaths and small ravines lead to charming transitions between pine and cotton, hill and dale; everywhere there is a gay medley of laurels, myrtles, cypresses, and above all, shining camellias.

The land is of volcanic origin, and the entire surface belongs to the tufa and the diluvium formation. The soil on the hills consists of an extremely fine, yet not over fat brown clay; whereas that of the valleys is throughout the country, with some trifling modifications, of a black, loose, and deep garden mould, which upon trial in different places I found extended to a depth of 12 to 15 feet, being throughout of the same quality, though somewhat more compact in the deeper layers. An impermeable stratum of clay probably underlies this arable crust. As the clay strata of the mountains, in consequence of the frequent and copious falls of rain, give rise to a multitude of springs, which are everywhere at hand, and may thus easily and without any great skill, be turned to account for the purpose of irrigation; so the impermeability of the stratum underlying the surface soil in the valleys enables the Japanese husbandmen to turn the soil at pleasure into a swamp, for the cultivation of rice.

Whichever way one may feel inclined to decide the question, whether the present fruitfulness of the soil is simply the artificial product of cultivation continued for a period of several thousand years, or whether this fertility existed from the beginning, making this people love and cherish the labours of agriculture, this much must be granted, at all events, that the clay of the diluvium, the mild climate, and abundance of water, afforded all the conditions, and the most convenient means, for a thriving cultivation. All these advantages have been most carefully turned to account by an industrious, ingenious, and sober people; and husbandry in Japan has become a truly national occupation. The Japanese have thoroughly mastered the difficult task of maintaining agriculture in a state of the highest perfection, although its pursuit is entirely in the hands of peasants and yeomen, who take rank in the sixth and last but one class of the social scale, and no Japanese gentleman is a farmer. There are no agricultural institutions for instruction in husbandry, no agricultural societies, no academies, no periodical press to spread the teachings of science. The son simply learns from the father; and as the father knows quite as much as his grandfather and great grandfather before him, so he pursues exactly the same system of husbandry as any other peasant in any other part of the empire; but it is a matter of perfect indifference where the young agriculturist learns his business. The young pupil in husbandry will always be able to master a certain small amount of information which the experience of ages has shown to be true, so that it may be looked upon as positive knowledge, and a sort of hereditary heirloom.

I must confess that I experienced a feeling of deep humiliation on many occasions, when with this simple knowledge, *and the safe and uncontested practical application of it in husbandry* before my eyes, I thought of home. We boast that we are a civilized nation; in our land men of the highest intellectual attainments devote their best energy to the improvement of agriculture; we have everywhere agricultural institutions and agricultural societies, chemical laboratories and model farms, to increase and diffuse the knowledge of husbandry. And yet how strange that, despite all this, we still go on disputing, often so vehemently and acrimoniously, about the first and most simple scientific principles of agriculture; and that those who earnestly search after truth are forced to admit the infinite smallness of their positive and undisputed knowledge! How strange also that even this trifling amount of positive knowledge has as yet found so little application in practice!

Among the great questions which still remain in dispute with us, whilst in Japan they have long since been settled in the laboratory of an experience extending over thousands of years, I must mention as the most important of all, that of manuring. The educated sensible farmer of the old world, who has insensibly come to look upon England, with its meadows, its enormous fodder production and immense herds of cattle, and in spite of these with its great consumption of guano, ground bones, and rape-cake, as the beau ideal and the only possible type of a truly rational system of husbandry, would certainly think it a most surprising circumstance to see a country even much better cultivated, without meadows, without fodder production, and even without a single head of cattle, either for draught or for fattening, and without the least supply of guano, ground bones, saltpetre, or rape-cake. This is Japan.

I cannot help smiling when I remember how, on my passing through England, one of the great leaders of agriculture in that country, pointing to his abundant stock of cattle, endeavoured with an authoritative air to impress upon my mind the following axioms, as the great secret of true wisdom:—'The more fodder, the more flesh; the more flesh, the more manure; the more manure, the more grain!' The Japanese peasant knows nothing of this chain of conclusions; he simply holds fast to one indisputable axiom, viz. without continuous manuring there can be no continuous production. A small portion of what I take from the soil is replaced by nature (the atmosphere and the rain), the remainder I must restore to the ground; the manner in which this is done is a matter of indifference. That the produce of the land has first to pass through the human body before it can be returned to the soil, is, as far as manuring is concerned, simply a necessary evil, which always involves a certain loss. As to the intermediate stage of cattle feeding, which we deem so requisite in our system, the Japanese farmer cannot at all see its necessity. He argues in his way that it must cost a great deal of unnecessary and expensive labour to have the produce of the field first eaten by cattle, so troublesome and expensive to breed, and that this system must involve more considerable loss of matter than his own. How much more simple it must be to eat the corn yourself, and to produce your own manure! Far from me be it, however, upon the ground of the so widely differing results to which the developement of agriculture has led in the two lands, to pass judgement upon our system of husbandry, and to exalt unduly that of the Japanese by attributing superior intelligence to that nation. Circumstances have brought about the results in question, and the following more especially have exercised a decided influence in the matter. The religious belief of the two great sects in Japan, the Sintoists and the Buddhists, forbids the eating of flesh, and not alone of flesh, but of everything derived from animals (milk, butter, cheese); this prohibition, of course, disposes of one of the principal objects for which cattle are bred. Even sheep, if kept for the wool alone, would not pay, as our farmers begin to find out even in Germany.

The very limited area of the homesteads in Japan also makes the maintaining of cattle superfluous. The smallness of the farms must not be attributed, however, to any excessive tendency to subdivision of landed property, but to the fact that the land belongs to the great princes or Daimios of the country, who have bestowed it in fee upon the lower nobility. The latter, again, being precluded by the institutions of the country from farming their own estates, have parcelled the land out, apparently from time immemorial, on perpetual leases, among the peasantry of the country. The size of these farms varies from two to five acres; the limitation having been most likely determined either by their natural position, or from the course of some brook or rivulet. Now, as this limited area is intersected moreover by drains and ditches, it will be readily seen that there is hardly a plot of ground to be found where the use of beasts of burden might be profitably had recourse to.

Now, with us matters are very different in these respects. We have a notion that we could not possibly exist in health and vigour without a considerable consumption of meat, although we have the fact constantly before our eyes, that our labourers, who assuredly require as much strength as any other class of society, are, for the most part, involuntary Buddhists. Our farms are always sufficiently large to preclude the notion of working them by hand, even leaving out of consideration the important circumstance that the price of labour is rather too high, in proportion to the value of the produce, to admit of such a system of farming. But that the culture of the soil is everywhere in the world in direct ratio to the division of the land is a well-established fact, of which the reality and significance are made most clearly apparent to the traveller who passes from the north of Germany to Japan, viâ England.

The only manure-producer, therefore, in Japan is man; and we need not wonder that the greatest care should be bestowed in that country upon the gathering, preparing, and applying his excrements. Now, as their entire course of proceeding contains much that is highly instructive for us, I consider it my duty to give as detailed a description of it as possible, even at the risk of offending the delicate feelings of the reader.

The Japanese does not construct his privy as we do in Germany, in some remote corner of the yard, with half-open rear, giving free admission to wind and rain; but he makes it an essential part of the interior of his dwelling. As he ignores altogether the notion of a 'seat,' the cabinet, which, as a general rule, is very clean, neat, and, in many cases, nicely papered or painted and varnished, has a simple hole of the shape of an oblong square running across and opposite to the entrance door, and serving to convey the excrements into the lower space. Squatting over this hole, with his legs astride, the Japanese satisfies the call of nature with the greatest cleanliness. I never saw a dirty cabinet in Japan, even in the dwelling of the very poorest peasant. It appears to me that there is something very practical in this form of construction of a closet. We, in Germany, construct privies over our dung-holes, and behind our barns, for the use of our farm-servants and labourers, and provide them with seats with round holes. With even only one aperture, it is too often found that after a few days' use they look more like pigstyes than closets for the use of man, and this simply because our labourers have a decided, perhaps natural, predilection for squatting. The construction of the Japanese privies shows how easy it would be to satisfy this predilection.

To receive the excrements, there is placed below the square hole a bucket or tub, of a size corresponding to it, with projecting ears, through which a pole can be passed to carry the vessel. In many instances a large earthen pot, with handles, is used, for the manufacture of which the Japanese clay supplies an excellent material. In some rare instances in the towns, I found a layer of chopped straw or chaff at the bottom of the vessel, and occasionally also interspersed among the excrements, a proceeding which, if I mistake not, has of late been recommended also in Germany. As soon as the vessel is full, it is taken out and emptied into one of the large dung-vessels. These are placed either in the yard or in the field. They are large casks or enormous stoneware jars, in capacity of from 8 to 12 cubic feet, let into the ground nearly to the brim. It is in these vessels that the manure is prepared for the field. The excrements are diluted with water, *no other addition of any kind being made to them*, and stirred until the entire mass is worked into a most intimately intermixed fine pap. In rainy weather, the vessel is covered with a moveable roof to shield it from the rain; in dry weather this is removed, to allow the action of the sun and wind. The solid ingredients of the pap gradually subside, and fermentation sets in; the water evaporates. By this time the vessel in the privy is again ready for emptying. A fresh quantity of water is added, the whole mass is again stirred and most intimately mixed together, in short, treated exactly like the first emptying. The same process is repeated, until the cask or pan is full. After the last supply of excrements, and thorough mixing, the mass is left, according to the state of the weather, for two or three weeks longer, or until it is required for use; *but under no circumstance is the manure ever employed in the fresh state.* THIS ENTIRE COURSE OF PROCEEDING CLEARLY SHOWS THAT THE JAPANESE ARE NO PARTISANS OF THE NITROGEN THEORY, AND THAT THEY ONLY CARE FOR THE SOLID INGREDIENTS OF THE DUNG. *They leave the ammonia exposed to decomposition by the action of the sun, and its volatilisation by the wind, but take the greater care to shield the solid ingredients from being wasted or swept away by rain, &c.* As the peasant, however, pays his rent to his landlord not in cash, but in a certain stipulated percentage of the produce of his fields, he argues quite logically that the supply of manure from his privy must necessarily be insufficient to prevent the gradual exhaustion of the soil of his farm; notwithstanding the marvellous richness of the latter, and in spite of the additional supply of manuring matter derived from the water of the brook or canal from which he takes his material for irrigation. He places, therefore, wherever his field is bordered by public roads, footpaths, &c., casks or pots buried in the ground nearly to the rim, urgently requesting the travelling public to make use of the same. To show how universally the economical value of manure is felt and appreciated in all classes of society in Japan, from the highest to the lowest, I need simply state the fact that, in all my wanderings through the country, even in the most remote valleys, and in the homesteads and cottages of the very poorest of the peasantry, I never could discover, even in the most secret and secluded corners, the least trace of human excrements. How very different with us, in Germany, where it may be seen lying about in every direction, even close to privies!

I need not mention that the manure thus left by benevolent travellers is treated exactly in the same way as the family manure.

But the excrements of the peasant contain also some other matter, which has not been derived from the soil of his fields, and which may be said to represent an additional importation of manure. The river, brooks, and canals, and the numerous little bays, abound in fish, which the religion of the Japanese permits him to eat, a permission of which he most largely avails himself. Fishes, crabs, lobsters, and snails are eaten in quantities, and these ultimately afford a most valuable item of contribution to the privy, and consequently to the fertilising field-manure.

The Japanese farmer prepares also *compost*. As he keeps no cattle to turn his straw, &c., into manure, he is forced to incorporate this part of his produce with the soil without 'animalisation.' The method pursued to effect this object consists simply in the concentration of the materials. Chaff, chopped straw, horse-dung, excrement gathered in the highways, tops and leaves of turnips, peelings of yams and sweet potatoes, and all the offal of the farm, are carefully mixed with a little mould, shovelled up in small pyramidal heaps, moistened and covered with a straw thatch. I often saw also in this compost heaps of shells of mussels and snails, with which most of the rivulets and brooks abound, and which, in all parts close to the seashore, may be obtained in any quantities. The compost heaps are occasionally moistened and turned with the shovel, and thus the process of decomposition proceeds rapidly, under the powerful action of the sun. I have also often seen the shorter process of reduction by fire resorted to when there was plenty of straw, or where the manure was required for use before it could be got ready by the fermentation process.

The half-charred mass was, in such cases, in so far as my own observation enabled me to judge, strewed directly on the seed sown in the ground.

I think the treatment of this compost is another proof that the Japanese farmer does not care for the azotised matters, and that he strives to destroy all organic substances in his manure before making use of it. *The great object of the Japanese farmer in all this is to turn his manure to account as promptly as possible.*

To attain this end, besides preparing his manures in the manner described, he has recourse also to the following means:—

1. He applies his manures, and particularly his chief manure derived from his privy, invariably as much as possible in the liquid form.

2. *He knows no other mode of manuring than that of top-dressing.*.

When he wishes to sow, the land is laid in furrows, in the way to be more fully described hereafter, and the seed is strewn by hand, and covered with a thin and even layer of compost, over which liquefied and very dilute privy manure is poured. The manure is diluted in the buckets in which it is carried from the preparing tubs or pots to the seed furrow, as this is the only way to ensure uniform intermixing of the materials. As this manure has fully fermented, it may without danger be brought into immediate contact with the seed, and thus materially assist the first radication.

It may be that this Japanese system of manuring cannot as yet be introduced into Europe in its integrity. But with such excellent results to show for their proceedings, we might surely take a few lessons from these old practical men, and employ them with such modifications as our social relations require. At all events we might adopt in principle the following:—

1. The greatest possible concentration of manures; which must necessarily lead also to a material reduction of cost. When I stated that the Japanese does not trouble himself about the azotised matters in his manures, and that his land is, notwithstanding, in a most flourishing state of culture, this is no proof, however, that it might *not even be better*, perhaps, to endeavour to fix the nitrogen too. If a more practical system can be devised, of which however I have my doubts, combining the advantage of both, so much the better! But till something better is discovered, we might surely adopt that which experience has proved to be good.

2. Top-dressing, which is of course necessarily connected with cultivation in drills or furrows.

3. Liquid manuring: not to the extravagant extent, however, in which it was sought to be carried out in England, but in accordance with the present condition of German agriculture.

4. *Manuring with every crop.*

The Japanese never cultivates a crop without manuring it, but he gives each crop or seed exactly as much and no more manure than is required for its full developement. *He does not care about enriching the soil for future crops.* What he demands is simply a full crop in return for each sowing. How often do we hear our farmers talk about this manure being preferable to that manure on account of its fertilising action being 'more lasting;' yet with all our wise provision for the future, how far are we now behind the Japanese, who seem to look always to the next harvest only! As they manure for each fresh crop, and the term 'fallow' in our acceptation is entirely unknown to them, they are forced to distribute their yearly production of manure equally over the entire area of their land, which can be accomplished only by sowing in drills or furrows, and by top-dressing.

The contrast between this rational system and the profuse application of our long straw manure over the whole surface of the field is truly glaring.

I may also add here that the manure in the Japanese towns is never artificially turned into guano or poudrette, but is sent every night and morning in its natural form into the country around, to return again after a time in the shape of beans or turnips. Thousands of boats may be seen early each morning laden with high heaps of buckets full of the precious stuff, which they carry from the canals in the cities to the country. These boats come and go with the regularity of the post; it must be admitted, however, that it is a species of martyrdom to be the conductor of a mailboat of this kind. In the evening long strings of coolies are met with on the road, who having in the morning carried the produce of the country to the town, are returning home each with two buckets of manure, not in a solid or concentrated form, but fresh from the privies. Caravans of packhorses, which often have brought manufactured articles (silk, oil, lacquered goods, &c.), a distance of 200 to 300 miles from the interior to the capital, are sent home again freighted with baskets or buckets of manure; in such cases, however, care is taken to select solid excrements.

Thus in Japanese agriculture we have before us the representation of a perfect circulation of the forces of nature: no link in the chain is ever lost, one is always interlaced with the other.

I cannot refrain here from drawing a parallel in this respect between the Japanese and our system. In our large farms we sell a portion of the productive power of our soil in the form of corn, turnips, or potatoes; but our carts which convey the products to the town or to the gates of the factory, bring back no compensation. One of the links of the chain is lost. There is another portion of our produce devoted to the feeding of large herds of cattle, of which a considerable amount is sent forth in the form of fat cattle, milk, butter, or wool; this again is never returned, and thus a second link of the chain is lost. Another small portion we and our labourers consume. This last portion at least *might* be turned to proper account, if we only knew, like the Japanese, to save and use it more carefully and wisely. Will any one venture to assert that the privy manure of our farms is of the least real importance? I verily believe that, under present circumstances, the privy manure of an estate of a thousand acres would be barely sufficient for half an acre of ground. There remains, then, from our present agricultural system, out of the entire productive power withdrawn by the crops from the soil, only that portion returned by our cattle, a small part indeed of the whole, if we take into consideration its bulk, and reflect in how concentrated a form we have disposed of the rest of that power in the shape of grain, milk, or wool.

It may be objected, I am quite aware, that it is strange that our system of keeping large stocks of cattle does succeed in leading to a high state of cultivation and abundant produce. I admit the fact, only let us ascertain first its true significance. It is, above all, necessary to settle about the true acceptation of the term 'culture.' If by 'culture' is meant the capability of the soil to give permanently high produce, by way of *real interest on the capital of the soil*, I must altogether deny that our farms (with perhaps a few exceptions), can properly be said to be in a satisfactory state of culture. But we have by excellent tillage and a peculiar method of manuring, put them in a condition to make the entire productive power of the soil available, and thus to give immediately full crops. It is not, however, the interest that we obtain in such crops, but the capital itself of the soil upon which we are drawing. The more largely our system enables us to draw upon this capital, the sooner it will come to an end. The term 'culture' applied to such a proceeding is a misnomer. The peculiar method of manuring alluded to consists merely in our endeavouring to feed the soil of our fields with the largest possible supply of azotised matter. Now, ammonia and the other azotised compounds may no doubt be looked upon as excellent agents to stir up the hidden and slumbering forces of the soil. But after all, these agents may be regarded somewhat in the light of a banker, who kindly exchanges the pound we have to spend for thirteen shillings; and then we can spend the change fast enough. This accounts for the large party amongst us who love and cherish the obliging banker.

This is the great difference between European and Japanese culture. The former is simply a delusion, which will be detected sooner or later. Japanese cultivation, on the other hand, is actual and genuine; the produce of the land represents indeed the interest of the capital of the soil's productive power. As the Japanese knows that he has to live upon that interest, his first care is devoted to keeping the capital intact. He only takes away from his soil with one hand, if he can make up the loss with the other; and he never takes more than he can return. He never endeavours to force the production by large supplies of azotised matters.

The fields in Japan do not, therefore, as a general rule, present that luxuriant aspect which gratifies our sight occasionally at home. There are no impenetrable forests of straw from six to eight feet high, to be seen, nor turnips weighing 100 lbs., with 99 lbs. of water in them. There is nothing extravagant in the sight of Japanese crops. *But what distinguishes them most favourably as compared to ours is their certainty and uniformity for thousands of years. The real produce of land can be calculated only by the average crops of a long number of years.*

If additional proof were needed to show that the state of cultivation is very superior, and that the land yields abundant produce, I would point to the fact that the Japanese empire, which covers an area similar to Great Britain and Ireland, and of which one-half at the most, from the hilly nature of the country, can be looked upon as fit for tillage, not only contains a larger number of inhabitants than Great Britain and Ireland, but maintains them without any supply of food from other parts. Whilst Great Britain is compelled to import corn from other countries, to the extent of many millions per annum, Japan since the opening of its ports actually exports no inconsiderable quantities of food.

SECTION II.

TILLAGE OF THE SOIL.

Deep cultivation of the soil has become a kind of proverb with our modern writers on agriculture; and the p.inciple of the system is, at least, fully admitted on all hands, the only objection occasionally raised against it being that it requires a large supply of manure. But the most enthusiastic admirer of the system in Europe can hardly conceive how universally and in what high perfection it is carried on in Japan.

The Japanese husbandman has come to treat his field as a plastic material, to be turned to account in any way or form he pleases, just as a tailor may cut out of a piece of cloth, cloaks, coats, trowsers or vests, and occasionally makes the one out of the other. To-day we find a plot of ground covered with a wheat crop; in eight days the wheat is reaped, and one half of the field is transformed into a swamp thoroughly saturated with water, in which the farmer, sinking up to his knees, is busy planting rice, whilst the other half is a broad and dry plot, raised 2 or 2½ feet above the rice swamp, and ready to receive cotton, or sweet potatoes, or buckwheat. It often happens also that a square plot in the centre is turned into a dry bed, surrounded by a broad rice swamp; and as the water must cover the surface of the latter only slightly, the levelling must have been effected with great care, and with the use of instruments.

The whole of this work has been done by the farmer and his small family in a very short time. That it could be accomplished in so short a time is a proof of the *great depth of the loose arable soil*, even after a harvest; and that the farmer could venture to do so without troubling himself about the next crop, is a sign of the *abounding wealth* of the soil in mineral constituents. It is only when great depth of the loose arable soil is combined with a plentiful store of mineral constituents that deep tillage of the ground can truly be resorted to. The description here given is not a mere fiction or creation of the imagination, but a faithful statement of facts such as I have had occasion to witness by the hundred. Considering that rice requires at least from 1 to 1½ feet of cultivated soil, and adding to this half the height of the raised bed, viz. 1 to 1¼ feet, this gives a cultivated depth of arable soil of from 2 to 3 feet.

This system of working the land at pleasure either as a raised dry plot or as a swamp, is indeed, at present, in Japan, simply a proof of the *existence* of deep tillage; but it is clearly evident that it must have been, at one time, also, the *means* of effecting it. If we are always to wait until we have collected a sufficient excess of manure (at the best but a very relative term), before proceeding to deepen the arable crust of our land, we may certainly predict that the system will but very rarely make any progress with us. Everybody knows that one cannot learn to swim without going into the water.

The introduction and constant progress of the system of deep tillage have been powerfully assisted in Japan by the practice pursued from time immemorial of growing all crops in drills. With the advantage of this method we have also long been familiar

Among the favourable features presented by the cultivation of root crops, our books of agriculture always place in a prominent rank the fact that it enables the farmer to deepen the arable soil of his land. All our gardeners, at least, have long ago adopted it.

I was not fully aware of the true importance of the method of growing crops in drills, until I had occasion to see it carried out to the fullest extent in Japan. We, in Europe, are as yet far from having adopted this plan as an essential part of our system of husbandry; we look upon the question still in a very *one-sided point of view, only in reference to the individual crop which we wish to grow.* But the Japanese farmer has raised it to the rank of a system, by which he has fully emancipated himself from the necessity of paying, as we are compelled to do, the least regard to the rotation of crops. By its means he has truly become master of his land. He has not only succeeded in growing crops at the same time which used to follow each other, but he has carried to the highest perfection the principle of mixed cultivation, which begins now to find favour also with our European farmers: he has, in this respect put an end to our confused and haphazard way of mixing crops on the same field, having by the adoption of the method of drill planting, brought order and regularity into the system. The following description of the Japanese system may serve by way of illustration.

We have a Japanese field before us, in the middle of October, with nothing but buckwheat upon it. The buckwheat is planted in rows, 24 to 26 inches apart; the intervening, now vacant, space had been sown in spring with small white turnip-radishes, which have already been gathered. These intervening vacant spaces are now tilled with the hoe to the greatest depth attainable by the implement. A portion of the fresh earth is raked from the middle up to the buckwheat, which is now in full flower: a furrow is thus formed in the middle, in which rape is sown, or the grey winter pea, the seed being manured in the manner already described, and seed and manure afterwards covered with a layer of earth. By the time the rape or the peas have grown one to two inches high, the buckwheat is ripe for cutting. A few days after the rows in which it stood are dug up, cleared, and sown with wheat or winter turnips. Thus crop follows crop the whole year through. The nature of the preceding crop is a matter of indifference, the selection of the succeeding one being determined by the store of manure, the season, and the requirements of the farm. If there is a deficiency of manure, the intervening rows are allowed to lie fallow, until a sufficient quantity has been collected for them.

This system, as a whole, has also this great advantage, that the manure may be used at all times, and need never lie idle as a dead capital bearing no interest; and moreover, perhaps, the most important point of all is, that a direct ratio is thereby secured between the power of the soil, as shown in the crops, and the stock of manure on hand, a ratio not disturbed here by artificial means or by any '*tour de force.*' Expressed in other words, the income and expenditure of the soil are always kept evenly balanced.

I have seen this system carried out to the fullest attainable degree in the vicinity of large towns, such as Jeddo, also in particularly fertile valleys, and on fields bordering on the great highways. Here crop succeeded crop, manure followed manure. Here the plot of ground produced much more than could be consumed on it; but the great city and the privies on the high road returned a supply of manure to balance the export of produce.

I have, however, also had occasion to visit farms situated on some hilly part far away from the high road, and only recently reclaimed and cultivated. As the Japanese farmer, as a general rule, prefers the valleys to the hilly ground, the supply of manure here is more restricted and more difficult, and any addition to it from towns or by travellers is almost altogether out of the question. Here I found occasionally only one crop on the ground; yet the rows were so wide asunder that another crop would have found ample space between them. With this system it is at least possible to till properly and repeatedly the intervening spaces, which are intended to receive the next crop; besides the constant supply of fresh earth to the present crop, by raking, places a larger store of soil at the disposal of the latter than could be done in any other way. In this manner only the one-half of the field (corresponding to the limited supply of manure) is actually made to produce; but the system of planting the crop in drills wide asunder always gives a much more abundant return than could possibly be obtained, if the one-half of the field as a continuous plot were completely sown, the other half being allowed to lie fallow. As the home production of manure or the importation of it from other parts, increases, the farmer proceeds to fill part also of the vacant rows, which thus leaves only the third or fourth part of the field fallow, until, at last, every row is made to produce crops.

How wide the difference between this system and ours! When we break up and till a plot of ground, we begin by extracting from it three or four harvests, without bestowing a particle of manure, and apply manure only when the soil is exhausted. *The Japanese husbandman never breaks up a plot of land, unless he possesses a small stock of manure*, which he may invest in the ground; and even then he only cultivates this new plot to the extent his supply of manure will permit. This rational proceeding shows the deepest insight into the nature of the system of agriculture to be pursued with a reasonable prospect of securing a constant succession of remunerative crops. No other illustration can so clearly show the difference between our European way of viewing the matter and the Japanese. We, in Europe, cut down the trees on a forest plot, sell the timber, grub up, plough and till the ground, and then proceed to dispose of the productive power of the new soil, in three cereal crops, obtained without the least supply of manure; or we may possibly assist in accelerating the exhaustion of the ground by a small dose of guano. All that this course of proceeding is calculated to accomplish is, that we have now to distribute the manure hitherto produced on our estate over a somewhat more extended surface than formerly. When the Japanese husbandman breaks up a plot of ground, he finds a virgin soil, the productive power of which he has not the least intention of impairing. He therefore, from the very outset, takes care to establish a proper balance between crop and manure, expenditure and income, maintaining thus intact the productive power of the ground, which is all that can reasonably be attempted by any rational husbandman ('Annal. der Preuss. Landwirthschaft,' January, 1862).

APPENDIX H. (page 237).

We would earnestly recommend all inquiring travellers in other parts of the world, to endeavour to ascertain, above all things, what are the proportions of the annual produce of the various cereals and cultivated plants raised in a continued succession of crops on unmanured soil of different kinds in the same place, and under the climatic influences of widely differing degrees of latitude. In so far as the author has been able to obtain reliable information on the matter, from various countries, more especially from the torrid zone, a careful examination of the facts ascertained would appear to refute everywhere the old widespread error that a very fruitful soil, under favourable climatic conditions, in the tropics for instance, will continue inexhaustible, even without receiving back from the hand of man the mineral matters removed in the crops. Even in the most enchanting lands of the tropical zone, on the most fruitful volcanic earth, such as is found in the old country of the Incas, the tableland of Quito, Imbabura, Riobamba, Cuenca, &c., a long-continued succession of crops drained the soil wherever it was impracticable to convey to the fields by artificial irrigation the mud carried down by the torrents of the Andes. In those regions water, aided by the widespread old volcanic mud streams (Lodozales), plays the part, which guano and farm-yard manure do elsewhere, of restoring to the soil the mineral constituents removed by a continued succession of crops. In most of the provinces of Persia, more especially in Aserbeidschan and in a great portion of Armenia and Asia Minor, the irrigation canals everywhere met with serve the purpose, not so much of moistening the ground, as of conveying to the land in the valleys the mineral detritus washed from the mountains at the time of the melting of the snow. This method of artificial manuring by irrigation is commonly applied also in those countries where there is no lack of rain and dew. It subserves the same purpose as the mud of the Nile in Egypt, viz. to replace the action of farm-yard manure. Where the mineral constituents removed by a long succession of crops are not restored to the ground either by animal manure, or by irrigation, the soil is almost completely drained of its productive powers, as is the case, for instance, in certain parts of the extensive tablelands of Tacunga and Ambato (in the South American State Ecuador), where barley will often barely give a two or threefold return, notwithstanding the frequent alternations of rain and sunshine. From the most reliable information obtained by me, even the most fertile estates in San Salvador and Chiriqui, in Central America, with their most fruitful, loose, trachytic soil, abounding in potash and silica, cannot show a single field on which maize has been grown for thirty years running without a considerable reduction of produce—a fact which sufficiently refutes the old mistaken notion of the inexhaustible fertility of the soil in the tropics.

On the western coast of Peru only those parts are extremely sterile, where no little artificial canals convey to the dry soil the water from the torrents of the Andes, which carries with it the mineral detritus washed from the declivities of the mountains. Wherever such artificial canals exist, and the conditions of the ground are favourable, the soil on the coast as well as in the interior of Peru and Bolivia is almost as productive as in the interior of the highlands of Ecuador, New Granada, and Gautemala. But it is not the water which is the agent in maintaining the steady productiveness of the soil, but, as in the case of the Delta of the Nile in Egypt, it is the mud carried along with the water, and which has been washed away from the disintegrated rocks of the Andes. The constituents of this mineral detritus, which are partly contained in the water in a state of minute mechanical division, and partly held in chemical solution, are brought to the fields by small channels. The water thus conveyed from the mountains in innumerable furrows is soon absorbed by the soil or evaporated, leaving a rich fertilising deposit behind. Pure rain water would be of very little avail, as, for instance, in the extensive tableland of Tacungar, with its barren pumice stone fields, where quite near the equator rain pours down almost daily during nine months of the year. It is not the atmospheric water that acts as the fertilising agent, but the muddy streamlets from the Andes. In Peru the fertilising action of guano is more enduring than in England, because the potash which the guano does not restore to the soil, is there supplied in the detritus from the trachytic constituents of the Andes ridge, which abound in felspar. This natural mineral manure is of the same high value in the South American lands of the Andes chain as the fertile *Löss*, accumulated by the great flood in past ages at the foot of the Bavarian and Swiss Alps. It is a fact full of meaning that the inhabitants of those parts of America should have arrived at the same simple means of restoring to the land the mineral constituents carried away by the crops, which are at the present day generally resorted to also under similar favourable conditions of the ground in the mountainous regions of Asia Minor, Armenia, Grusia, Western Persia, as well as in the north of Mesopotamia (Mossul), and, if I mistake not, in Thibet also. The waters of the rivers Kur, Araxes, Euphrates and Tigris, are in spring just as turbid and as much impregnated with mud, which simply means earthy particles, as the Nile, and as the East Persian river Herirud, which it is well known is altogether absorbed up in fields and gardens. The experience of ages past has no doubt taught the inhabitants of these ancient countries, in both hemispheres, this way of restoring to their fields the incombustible constituents removed from them in the produce carried away to the large towns (Professor Dr. Moritz Wagner; see supplement to 'Augsb. Allg. Zeitung,' No. 36, February 5, and No. 173, June 22, 1862).

APPENDIX I (page 321).

ANALYSIS OF CLOVER MADE BY DR. PINCUS.

100 parts of air-dried clover contained,—

	Unmanured.				Manured with sulphate of magnesia.				Manured with sulphate of lime.			
	Stems	Leaves	Flowers	Entire plant	Stems	Leaves	Flowers	Entire plant	Stems	Leaves	Flowers	Entire plant
Water	12·25	13·04	15·05	12·95	13·00	14·45	12·12	13·27	11·85	10·70	12·24	11·60
Vegetable fibre	39·55	15·07	16·36	28·85	39·47	12·58	17·08	29·70	38·75	13·73	16·96	29·87
Mineral constituents	5·05	11·16	6·32	6·95	6·75	10·97	7·47	7·94	6·65	11·45	7·45	7·96
Protein substances	10·15	22·08	17·59	14·70	11·42	24·37	19·59	15·81	12·34	28·74	20·57	17·45
Hydrate of carbon	33·00	38·65	44·68	36·55	29·36	37·63	43·74	33·28	30·41	35·38	42·78	33·12
	100·00	100·00	100·00	100·00	100·00	100·00	100·00	100·00	100·00	100·00	100·00	100·00
Total quantity of nutritive substances	43·15	60·73	62·27	51·25	40·78	62·00	63·33	49·09	42·75	64·12	63·35	50·57
Proportion of the protein substance to the hydrate of carbon	1 : 3·25	1 : 1·75	1 : 2·54	1 : 2·46	1 : 2·57	1:1·54	1:2 23	1:2·10	1:2·46	1:1·23	1:2·08	1:1·90

ASH CONSTITUENTS.

100 parts of ash contained,—

	Clover unmanured.	Clover manured with sulphate of magnesia.	Clover manured with sulphate of lime.
Chlorine	1·93	1·22	1·73
Carbonic acid	21·43	21·75	19·17
Sulphuric acid	1·33	2·36	3·29
Phosphoric acid	7·97	8·49	8·87
Silicic acid	2·67	2·55	3·08
Potash	33·58	32·91	35·37
Soda	2·12	3·03	2·73
Lime	21·71	20·66	19·17
Magnesia	5·87	5·27	5·47
Sesquioxide of iron	0·94	1·22	0·94
	99·55	99·46	99·82

CALCULATED UPON THE ASH FREE FROM CARBONIC ACID.

	Clover unmanured.	Clover manured with sulphate of magnesia.	Clover manured with sulphate of lime.
Chlorine	2·46	1.56	2·14
Sulphuric acid	1·69	3·02	4·07
Phosphoric acid	10·14	10·85	10·97
Silicic acid	3·40	3·26	3·81
Potash	42·73	42·05	43·77
Soda	2·70	3·87	3·37
Lime	27·62	26·40	23·72
Magnesia	7·47	6·74	6·77
Sesquioxide of iron	1·20	1·56	1·16
	99·41	99·31	99·78

The remarkable investigations by Dr. Grouven of the *clover disease* deserve also a place here.

The so-called 'clover disease' manifests itself in the clover plant, at the period of flowering, by the appearance of a multitude of brown spots of cryptogamic plants covering stems and leaves. The result of the affection is not simply a failure of the clover crop, but the produce reaped is unwholesome for cattle.

In his examination of the diseased clover, Grouven compared the organic and the ash constituents of the diseased with those of the healthy plant. Both the healthy and the diseased clover were produced from a mixture of seeds of red clover, lucerne, and esparsette, such as is usually grown at Salzmunde, where the experiments were made. The samples for examination and analysis were taken from the field on August 12. The analysis of the healthy plant was confined to the determination of the organic substances and the amount of ash.

100 parts of air-dried clover-hay contained,—

	Diseased clover.	Healthy clover.
Water	16·2	16·2
Protein substances	16·7	11·7
Fat	3·6	2·8
Saccharine matter, calculated as starch*	7·0	18·5
Non-azotised compounds unknown	17·9	11·3
Woody fibre	31·7	31·4†
Ash	6·9	8·1
	100·0	100·0

The composition of the ash of the diseased clover was compared with that of the ash of red clover (Wolff) and esparsette (Way).‡ The ashes were calculated after deduction of carbonic acid, sand, clay, and sesquioxide of iron.

	Diseased clover. (Grouven.)	Red clover. (Wolff.)	Esparsette. (Way.)
Potash	3·32	35·5	35·8
Soda	0·87	0·7	3·5
Lime	55·71	32·8	35·9
Magnesia	13·08	8·4	5·5
Chlorine	2·76	3·5	2·0
Sulphuric acid	13·46	3·3	2·8
Phosphoric acid	5·99	8·4	9·6
Silicic acid	4·88	7·0	4·3
	200·07	99·6	99·4

Grouven is led to conclude from the result of his examination that the primary cause of the clover disease is attributable to a

* Substances convertible into sugar by sulphuric acid.
† With 0·1 of ash and 0·184 of protein substances.
‡ Compare also the preceding analysis by Dr. Pincus.

change in the chemical composition of the plant, which again is caused by an altered condition of the soil. The very considerable deficiency of phosphoric acid and potash in the ash of the diseased plant is certainly remarkable ('Zeitschrift der landwirthschaftlichen Centralvereins der Provinz Sachsen, 1861,' page 73).

INDEX.

Use and Abuse
of
America's Natural Resources

An Arno Press Collection

Ayres, Quincy Claude. **Soil Erosion and Its Control.** 1936

Barger, Harold and Sam H. Schurr. **The Mining Industries, 1899–1939.** 1944

Carman, Harry J., editor. **Jesse Buel:** Agricultural Reformer. 1947

Circular from the General Land Office Showing the Manner of Proceeding to Obtain Title to Public Lands. 1899

Fernow, Bernhard E. **Economics of Forestry.** 1902

Gannett, Henry, editor. **Report of the National Conservation Commission, February 1909.** Three volumes. 1909

Giddens, Paul H. **The Birth of the Oil Industry.** 1938

Greeley, William B. **Forests and Men.** 1951

Hornaday, William T. **Wild Life Conservation in Theory and Practice.** 1914

Ise, John. **The United States Forest Policy.** 1920

Ise, John. **The United States Oil Policy.** 1928

James, Harlean. **Romance of the National Parks.** 1939

Kemper, J. P. **Rebellious River.** 1949

Kinney, J. P **The Development of Forest Law in America.** *Including,* Forest Legislation in America Prior to March 4, 1789. 1917

Larson, Agnes M. **History of the White Pine Industry in Minnesota.** 1949

Liebig, Justus, von. **The Natural Lawss of Husbandry.** 1863

Lindley, Curtis H. **A Treatise on the American Law Relating to Mines and Mineral Lands.** Two volumes. 2nd edition. 1903

Lokken, Roscoe L. **Iowa**—Public Land Disposal. 1942

McGee, W. J., editor. **Proceedings of a Conference of Governors in the White House, May 13–15, 1908.** 1909

Mead, Elwood. **Irrigation Institutions.** 1903

Moreell, Ben. **Our Nation's Water Resources**—Policies and Politics. 1956

Murphy, Blakely M., editor. **Conservation of Oil & Gas:** A Legal History, 1948. 1949

Newell, Frederick Haynes. **Water Resources:** Present and Future Uses. 1920.

Nimmo, Joseph, Jr. **Report in Regard to the Range and Ranch Cattle Business of the United States.** 1885

Nixon, Edgar B., editor. **Franklin D. Roosevelt & Conservation, 1911–1945.** Two volumes. 1957

Peffer, E. Louise. **The Closing of the Public Domain.** 1951

Preliminary Report of the Inland Waterways Commission. 60th Congress, 1st Session, Senate Document No. 325. 1908

Puter, S. A. D. & Horace Stevens. **Looters of the Public Domain.** 1908

Record, Samuel J. & Robert W. Hess. **Timbers of the New World.** 1943

Report of the Public Lands Commission, with Appendix. 58th Congress, 3d Session, Senate Document No. 189. 1905

Report of the Public Lands Commission, Created by the Act of March 3, 1879. 46th Congress, 2d Session, House of Representatives Ex. Doc. No. 46. 1880

Resources for Freedom: A Report to the President by The President's Materials Policy Commission, Volumes I and IV. 1952. Two volumes in one.

Schoolcraft, Henry R. **A View of the Lead Mines of Missouri.** 1819

Supplementary Report of the Land Planning Committee to the National Resources Board, 1935–1942

Thompson, John Giffin. **The Rise and Decline of the Wheat Growing Industry in Wisconsin** (Reprinted from *Bulletin of the University of Wisconsin,* No. 292). 1909

Timmons, John F. & William G. Murray, editors. **Land Problems and Policies.** 1950

U.S. Department of Agriculture—Forest Service. **Timber Resources for America's Future:** Forest Resource Report No. 14. 1958

U.S. Department of Agriculture—Soil Conservation Service and Forest Service. **Headwaters Control and Use.** 1937

U.S. Department of Commerce and Labor—Bureau of Corporations. **The Lumber Industry,** Parts I, II, & III. 1913/1914

U.S. Department of the Interior. **Hearings before the Secretary of the Interior on Leasing of Oil Lands.** 1906

Whitaker, J. Russell & Edward A. Ackerman. **American Resources:** Their Management and Conservation. 1951